Impacts of Violent Conflicts on Resource Control and Sustainability

Esther Akumbo Nyam
Plateau State University, Nigeria

Asiru Hameed Tunde
Umaru Musa Yar'adua University, Nigeria

A volume in the Practice, Progress, and Proficiency in Sustainability (PPPS) Book Series

IGI Global
DISSEMINATOR OF KNOWLEDGE

Published in the United States of America by
IGI Global
Engineering Science Reference (an imprint of IGI Global)
701 E. Chocolate Avenue
Hershey PA, USA 17033
Tel: 717-533-8845
Fax: 717-533-8661
E-mail: cust@igi-global.com
Web site: http://www.igi-global.com

Library of Congress Cataloging-in-Publication Data

Names: Nyam, Esther Akumbo, 1968- editor. | Tunde, Asiru Hameed, 1982- editor.
Title: Impacts of violent conflicts on resource control and sustainability / Esther Akumbo Nyam and Asiru Hameed Tunde, editors.
Description: Hershey, PA : Engineering Science Reference, [2019]
Identifiers: LCCN 2017061475| ISBN 9781522559870 (hardcover) | ISBN 9781522559887 (ebook)
Subjects: LCSH: War--Environmental aspects. | War--Economic aspects. | Mass media and war.
Classification: LCC TD195.W29 I47 2019 | DDC 333.7--dc23 LC record available at https://lccn.loc.gov/2017061475

This book is published in the IGI Global book series Practice, Progress, and Proficiency in Sustainability (PPPS) (ISSN: 2330-3271; eISSN: 2330-328X)

British Cataloguing in Publication Data
A Cataloguing in Publication record for this book is available from the British Library.

All work contributed to this book is new, previously-unpublished material.
The views expressed in this book are those of the authors, but not necessarily of the publisher.

For electronic access to this publication, please contact: eresources@igi-global.com.

Practice, Progress, and Proficiency in Sustainability (PPPS) Book Series

ISSN:2330-3271
EISSN:2330-328X

Editor-in-Chief: Ayman Batisha, International Sustainability Institute, Egypt

MISSION

In a world where traditional business practices are reconsidered and economic activity is performed in a global context, new areas of economic developments are recognized as the key enablers of wealth and income production. This knowledge of information technologies provides infrastructures, systems, and services towards sustainable development.

The **Practices, Progress, and Proficiency in Sustainability (PPPS) Book Series** focuses on the local and global challenges, business opportunities, and societal needs surrounding international collaboration and sustainable development of technology. This series brings together academics, researchers, entrepreneurs, policy makers and government officers aiming to contribute to the progress and proficiency in sustainability.

COVERAGE

- Intellectual Capital
- Innovation Networks
- Eco-Innovation
- Socio-Economic
- E-Development
- Technological learning
- Sustainable Development
- ICT and knowledge for development
- Environmental informatics
- Global Content and Knowledge Repositories

The Practice, Progress, and Proficiency in Sustainability (PPPS) Book Series (ISSN 2330-3271) is published by IGI Global, 701 E. Chocolate Avenue, Hershey, PA 17033-1240, USA, www.igi-global.com. This series is composed of titles available for purchase individually; each title is edited to be contextually exclusive from any other title within the series. For pricing and ordering information please visit http://www.igi-global.com/book-series/practice-progress-proficiency-sustainability/73810. Postmaster: Send all address changes to above address.

Titles in this Series

For a list of additional titles in this series, please visit:
https://www.igi-global.com/book-series/practice-progress-proficiency-sustainability/73810

Driving Green Consumerism Through Strategic Sustainability Marketing
Farzana Quoquab (Universiti Teknologi Malaysia, Malaysia) Ramayah Thurasamy (Universiti Sains Malaysia, Malaysia) and Jihad Mohammad (Universiti Teknologi Malaysia, alaysia)
Business Science Reference • ©2018 • 301pp • H/C (ISBN: 9781522529125) • US $215.00

Entrepreneurship and Business Development in the Renewable Energy Sector
Adrian Dumitru Tantau (Bucharest University of Economic Studies, Romania) and Laurenţiu Cătălin Frăţilă (Bucharest University of Economic Studies,Romania)
Business Science Reference • ©2018 • 381pp • H/C (ISBN: 9781522536253) • US $245.00

Technological Tools for Value-Based Sustainable Relationships in Health Emerging ...
Mosad Zineldin (Linnaeus University, Sweden) and Valentina Vasicheva (Linnaeus University, Sweden)
Medical Information Science Reference • ©2018 • 149pp • H/C (ISBN: 9781522540915) • US $155.00

Sustainable Local Energy Planning and Decision Making Emerging Research and ...
Vangelis Marinakis (National Technical University of Athens, Greece)
Information Science Reference • ©2017 • 171pp • H/C (ISBN: 9781522522867) • US $140.00

Sustainable Potato Production and the Impact of Climate Change
Sunil Londhe (International Centre for Research in Agroforestry (ICRAF), India)
Information Science Reference • ©2017 • 323pp • H/C (ISBN: 9781522517153) • US $180.00

Handbook of Research on Green Economic Development Initiatives and Strategies
M. Mustafa Erdoğdu (Marmara University, Turkey) Thankom Arun (University of Essex, UK) and Imran Habib Ahmad (Global Green Growth Institute, South Korea)
Business Science Reference • ©2016 • 747pp • H/C (ISBN: 9781522504405) • US $335.00

For an entire list of titles in this series, please visit:
https://www.igi-global.com/book-series/practice-progress-proficiency-sustainability/73810

IGI Global
DISSEMINATOR OF KNOWLEDGE

701 East Chocolate Avenue, Hershey, PA 17033, USA
Tel: 717-533-8845 x100 • Fax: 717-533-8661
E-Mail: cust@igi-global.com • www.igi-global.com

Editorial Advisory Board

Table of Contents

Detailed Table of Contents

This chapter explores the media coverage of violent conflicts and climate change issues in Nigeria from the perspective of human rights journalism. Nigeria has had a chequered history of violent conflicts, especially since it achieved self-rule from Britain in 1960. These conflicts have been wide-ranging and cross-cutting (political, ethnic, religious, communal, among others). The Nigerian media have equally had a long history of the coverage of these violent eruptions across the country. However, the mainstream media's approach to the coverage of these issues have tended to be on the physical coverage of the conflicts to, almost, neglect of some of the underlying causative factors, such as climate change. The phenomenon of climate change globally is significant, especially in developing countries, such as Nigeria. In recent times, the threats that are posed by climate change have been enormous. Some communities across the country have been sacked by its devastations, especially from the norther corridors of the country. This has therefore necessitated the forced migration of some of these communities, especially the Fulani herdsmen and their cattle to other parts of the country for greener pasture. This situation always resulted in conflict, which is often violent. This chapter argues that, although the federal government of Nigeria through its Ministry of Environment have come out with programs and policies/initiatives aimed at combating the menace of this phenomenon, and the media, apart from the fact that they have not fully keyed into these programs to step them down for the audience, there is the urgent need for them to review their approaches toward the coverage of these conflicts. This chapter advocates a human-rights-journalism-based approach to the coverage of these conflicts because of its diagnostic approach, which gives a critical reflection of the experiences of the victims of human rights violations in all its ramifications.

Climate change is a serious global issue and concern that is attributed to change. A change of climate that is directly or indirectly related to human activity, that which alters the composition of the global atmosphere and which in addition to natural climate variability observed over comparable time periods. There is therefore no doubt that the earth is warming, and the climate changing. Despotism and the rule of despots as agents of democracy has created a rift in the issue of climate change on its citizenry in the southern zone of Plateau State in the area of health, water shortages, cutting meals due to the economic recession in Nigeria. Research has shown that climate change can create a conflict, and it does have a direct effect on scarce resources required to sustain life. Water is at the heart of human existence. Global warming has a major impact on global water cycle, hence on rainfall, soil moisture, rivers, and sea levels. If climate change is not tackled urgently, the calamity will be enormous.

The study investigated the impact of palm kernel cake (PKC) in resolving herdsmen and arable farmers' violent conflicts in Ogun State, Nigeria. There is mounting evidence of violent conflicts, killings, and raping by herdsmen attacks in Ogun State communities. The study was carried out to examine the level experiences of violent conflicts, losses, gains (if any), and the impact of PKC in resolving further violent conflicts among arable farmers and herdsmen. A total of 215 registered poultry farmers were purposively selected which consisted of all the local government areas in Ogun State. A 40-item questionnaire was designed and used to collect information from the respondents. The data collected were analyzed using descriptive statistics such as frequency counts and percentages while inferential statistics such as chi-square were used. Findings revealed that there was significant relationship between the socio-economic characteristics of the respondents and their level experiences of violent conflicts.

Petro Trad's Lebanese republic was a transitory dominion entity that resulted from and was shaped by the friendly and allied France. The Lebanese-specific cosmopolitan entity was intended to become a Monaco-like structure with a unique, ethnically diverse financial system, with the Lebanese constitution developed to privilege a predominant ultraliberal banking and finance. The idea of a strong financial Franco-Lebanese entity melted down in front of a ruthless ethno-religious antagonistic conflict between Maronite Christians and the Muslim narrow majority that was also fueled by regional and global superpowers. Another very important factor that accelerated the meltdown was the attachment of remote rural and desert regions to the historical Lebanon, or Petro Trad's small Lebanon. The chapter explores the business transformation and enterprise architecture framework to analyze this change.

Filmmaking in Nigeria has no doubt attracted so much attention worldwide and the industry has become a major employer of labor and a source of national revenue. It has also given birth to regional language and ethnic film industries, including Kannywood and Yoruwood. These industries have become a voice for the voiceless, a medium to express cultures, values, and socio-political issues. However, the process of movie making in Nigerian conflict areas has been marred by seemingly sustained ethno-religious conflicts and attacks by insurgents in major film release points like Jos, Lagos, Kano, Kaduna, Aba, Onitsha, and Abuja. This discourse is informed by the need to interrogate how conflicts are detrimental to collective development of the Nigerian state and the film industry in particular. The Jos experience has generated concerns in the industry, hence the need to highlight reason(s) for the drop in film production on the plateau.

Conflict and war are not unique to Nigeria or Africa. It is now a truism that in conflict situations women are one of the worst hit. The reason is not farfetched: women are in conflict areas, and they are likely to be raped, kidnapped, killed, and or end up

groups in Southern Katsina emirate was associated to high activities of cattle rustlers and banditry by the Fulani pastoralist. These activities were identified to be the causes of armed violent conflict associated with farmer and herdsmen conflict. Thus, to underscore the nature of conflicts between the pastoralist and farmers in the area, the study has adopted structuration approach in describing the role of structure and individual agents in the conflict.

Foreword

It is projected that more than one billion people, including about 400 million of the world's acutely underprivileged, live in fragile nations, with severe poverty, and devoid of dependable and steady basic governance and security. It is estimated that more than half of the world's poorest people live in countries that are experiencing, or in danger of, violent conflict. Veritably, conflict hits the most vulnerable the hardest.

Access to natural resources is a common raison d'être for engaging in or fuelling conflict. These clashes habitually materialize for the reason that people have dissimilar uses for resources such as forests, water, pastures and land, or desire to administer them in dissimilar ways. Disputes also arise when these interests and needs are discordant, or when the priorities of some user groups are not considered in policies, programmes and projects. Wrangles with ethnicity or religion at their core can dovetail into dreadful atrocities and rapidly escalate into genocide. A race for political power can spark off a conflict, as can a struggle for scarce water and other valuable resources. Political, economic and social inequalities between groups engender antipathy and enable leaders to marshal and pit one group against another more easily. Such conflicts of interest are inevitable features of all societies. Violent conflict over land has occurred in many nations of the world: China, East Timor, Kosovo, Rwanda, Ethiopia, Kenya, and Tajikistan, among many others. It is estimated that close to seventy armed conflicts active in recent times have been closely linked to natural resources exploitation, in which either official or illegal exploitation aided to elicit, intensify, or protract bloodshed. In some parts of the world violent protests have broken out over the distribution of water. In the Middle East, disputes over oil fields, among other issues, led to the first Gulf War.

Whilst probing the diverse types of conflicts, sources thereof and representation of conflict in Nigeria, this book also appositely, ornately and nimbly dissects the impact of these conflicts, averring that if these clashes are not addressed, they may stimulate environmental degradation, disrupt projects and undermine livelihoods. The authors avow that societies in violent conflict strain to supply even the most rudimentary of services such as water, security and education as their economies constrict. Millions of people are left with little to eat, no access to health care or education; and many become refugees. Communities struggle to survive as food and fuel prices put jobs and lives in jeopardy

Acknowledging that conflict is a common feature of any resource use system is a prerequisite for sustainable management that is participatory and equitable. Through a multidisciplinary slant, this book not only offers a panoramic outlook of the disposition and foundation of various violent conflicts in Nigeria, but also suggests ways of resolving conflicts, building sustainable peace and subsequently providing more effective, faster and more transparent support for immediate post-conflict needs. The researchers recorded herein have done a magnificent job in identifying gaps in our knowledge and understanding.

John Mugubi
Kenyatta University, Kenya

Preface

The spate of violent conflicts in developing countries in recent years has attracted concerns from scholars from all fields of study. Conflicts generally have the tendency of plunging societies into disarray. Indeed, the experience of the last century and a decade and a half into the millennium has thrown some countries into economic recession while other countries have also disintegrated politically and socially. In fact, countries that have been doing well economically prior to eruption of violent conflicts were left with no option but to watch their years of painstaking development efforts being swept away, leaving untold hardships that may linger for decades. In regions where there are natural resources, such as minerals, oil, gold, diamond, timber, etc. violent conflicts may be inevitable owing to the absence of a level playing field as well as inequalities. This could pose challenges on the resource control and sustainability. This shows that there is a somewhat complex connection between natural resources and violent conflicts and is not unconnected to the modus operandi of the exploitation of the resources. Thus when natural resources are exploited, either legally or illegally, there are attendant beneficiaries and those who would be affected adversely. As a corollary to that, the exploitation brews or sustains violent conflicts.

Conflicts are also exacerbated by the current wave of climatic threats all around the globe. Many regions are experiencing ecological consequences such as land degradation such that they cannot have access to portable drinking water because of contamination of water sources and resources, destruction of farmlands and dispersion of toxic substances. What the people usually resort to is militancy and rebellion against the exploiters; for instance, the case of the Niger Delta militants in Nigeria who took up arms in protest against their land degradation, poverty and call for fair distribution of their resources. A similar case is the Fulani-Herdsmen crisis in Benue and some other parts of Nigeria where farmers and indigenes of the affected states complain bitterly of being attacked by Fulani herdsmen on their account of disallowing their cattle from grazing on their farmland. In this book, contributors, mainly from Nigeria, have examined violent conflict from different perspectives, such as preventive measures in stemming the tide of violent conflict, media reportage in conflict situation, women and children in conflict situations, conflict and barriers etc.

Similarly, the abundance of national resources in some regions often leads to government mismanagement or weak policy framework which only fuels violent conflicts. While it can be argued that all violent conflicts are politically motivated, in a particular sense, political violence manifests not only in a given political dispensation but at other levels of relationships. For example, the social, cultural and even linguistic terrain continue to invite all sorts of players into the political space and inevitably leads to contestations. The interconnection between resource control, violent conflicts and politics is the glaring fact that without resources, it would be difficult to have a stable political atmosphere.

This book therefore, timely as it is, unearths the various dimensions of violent conflicts, how they have been managed or otherwise and how such practices and efforts contribute to resource control and sustainability. By extension, this collection adds to scholarship in the areas of conflicts, resource control and sustainability, and climate change which are undoubtedly global issues that many countries are seriously and frantically concerned about.

In recognition of the above, this book collection, *Impacts of Violent Conflicts on Resource Control and Sustainability*, could not have come at a better time than now when violent conflicts arising from political, climate, religious, ethnic etc influences are being experienced from different countries of the world. In this collection, therefore, one of the perspectives to solving violent conflict is the perspective of Theatre for Global Development (TFGD). This approach is applied to the southern zone as enlightenment in creating conscientization, awareness and participation on water harvest for sustainable results. The chapter explores the assumption that Theatre for Global Development and its techniques are appropriate instruments for enlightenment and improved participatory actions in Jos, Plateu Nigeria.

From economic point of view, Abiala's chapter explored the impact of Palm Kernel Cake (PKC) in resolving violent conflict among herdsmen and farmers in Ogun State, Nigeria. The chapter advocates for the use of palm kernel cake as alternative or supplementary to pasture grasses or weeds as practiced by other developing nations, such as Malaysia and Indonesia in order to avoid violent conflicts among herdsmen and farmers.

From media perspective, there is an assessment of the media coverage of the spate of violent conflicts against the backdrop of contemporary challenges in Nigeria and the paper argues that, although violent conflicts are still rife and wide-ranging, natural resources, such as land and oil, appear to be uniting adversaries today. The case of the Urhobo and Isekiri of the Niger-Delta region of the country is a testimony where they reconciled their differences and now formed a common front against the Nigerian state in their agitation for a greater control of the oil resource.

Similarly, the diverse ethnic groups of Plateau State in north central Nigeria are equally resolute in their opposition to the federal government's proposed policy on 'grazing reserve', which is seen as an insidious avenue to cede their lands to the Fulani; the Southern Kaduna ethnic groups in south-eastern part of the state (Samuel, 2012) appear to be speaking with one voice against the militant Fulani herdsmen whose reign of terror is perceived to be another clandestine way to dispossess the natives of their lands. On stemming the tide of violent conflicts, media must be proactive rather than been reactive by focusing on deconstructing the embedded structural causes of violence, such as forced migration, climate change, Fulani herdsmen onslaught, insecurity, and exclusion of minorities.

In a related dimension to TFD, another chapter explores the effect of violent conflicts on the process of movie making, especially in Nigerian major film release points like Jos, Lagos, Kano, Kaduna, Aba, Onitsha and Abuja. This discourse is informed by the need to interrogate how conflicts are detrimental to collective development of the Nigerian state and the film industry in particular. The Jos experience has generated concerns in the industry because during the heat of the crises (2007-2011), film production and economic activities dropped drastically.

From literary perspective, another chapter explores the plight of women in conflict situations both within the text and the context that is Nigeria- using Femi Osofisan's "Women of Owu" (an adaptation of Euripedes' "Trojan Women"). The paper argues that the manner in which Osofisan dramatizes the story of the African woman is both sympathetic and resilient, yet a gruesome reminder of the disadvantageous position women find themselves in conflict situations.

Finally, from Lebanon, Trad explores how the idea of a strong financial Franco–Lebanese entity melted down in front of a ruthless ethno-religious antagonistic conflict between Maronite Christians and the Muslim narrow majority that was also fuelled by regional and global superpowers. The chapter uses historical financial and geopolitical crisis to present a recurrent and frequent scenario of a financial destruction model based on a brutal and sudden societal transformation event that Taleb would call a black swan. The author uses the case of the Lebanon in an attempt to demonstrate this research's mathematical modelling method and analyzing tool.

This book will go a long way in benefiting researchers who are interested in exploring and understanding various perspectives on causes and management of violent conflicts. It will be useful to policy makers, government functionaries, peace keepers and non-governmental organizations in making informed decisions, mediation and provision of relief resources in cases of violent conflicts and to stem the tide of violent conflicts of various forms and ensure security of lives and properties. To academicians and the general public, the book will serve as a good reference for enhancing a peaceful and violent free world, and how to manage resources during conflicts.

Acknowledgment

We, the editors, wish to thank our contributors, who despite their schedules had to cope with our numerous mails and 'intimidating' deadlines. We appreciate your cooperation and understanding from the outset of this project, especially during the peer review phase; these are marks of scholarly attitude. We, therefore, hope that this collection will enhance more scholarly working relationships. We thank other colleagues who did not contribute a chapter to this volume but assisted by offering insightful and enriching comments during the reviews.

We also wish to thank Dr. John Mugubi of the School of Creative Arts, Film and Media Studies, Kenyatta University for writing a foreword for this collection.

Esther Akumbo Nyam
Plateau State University, Nigeria

Asiru Hameed Tunde
Umaru Musa Yar'adua University, Nigeria

Chapter 1
Fulani Herdsmen and Farmers Conflicts in Nigeria:
Approaches to Media Coverage of Conflict and Climate Change Issues

Andrew Danjuma Dewan
University of Salford, UK

ABSTRACT

This chapter explores the media coverage of violent conflicts and climate change issues in Nigeria from the perspective of human rights journalism. Nigeria has had a chequered history of violent conflicts, especially since it achieved self-rule from Britain in 1960. These conflicts have been wide-ranging and cross-cutting (political, ethnic, religious, communal, among others). The Nigerian media have equally had a long history of the coverage of these violent eruptions across the country. However, the mainstream media's approach to the coverage of these issues have tended to be on the physical coverage of the conflicts to, almost, neglect of some of the underlying causative factors, such as climate change. The phenomenon of climate change globally is significant, especially in developing countries, such as Nigeria. In recent times, the threats that are posed by climate change have been enormous. Some communities across the country have been sacked by its devastations, especially from the norther corridors of the country. This has therefore necessitated the forced migration of some of these communities, especially the Fulani herdsmen and their cattle to other

DOI: 10.4018/978-1-5225-5987-0.ch001

parts of the country for greener pasture. This situation always resulted in conflict, which is often violent. This chapter argues that, although the federal government of Nigeria through its Ministry of Environment have come out with programs and policies/initiatives aimed at combating the menace of this phenomenon, and the media, apart from the fact that they have not fully keyed into these programs to step them down for the audience, there is the urgent need for them to review their approaches toward the coverage of these conflicts. This chapter advocates a human-rights-journalism-based approach to the coverage of these conflicts because of its diagnostic approach, which gives a critical reflection of the experiences of the victims of human rights violations in all its ramifications.

INTRODUCTION

This chapter analyses the media coverage of the spate of violent conflicts against the backdrop of the Fulani herdsmen/farmers in Nigeria. Although violent conflicts are still rife and wide-ranging, natural resources, such as land is a major cause of conflict. The chapter asserts that, while much attention is being given to the media coverage of conflicts caused by the struggle for land, the challenges that climate change pose are not being given equal press coverage. The United Nations sponsored climate change conference in Paris in 2015, which brought together world leaders and governments of about 190 nations was not only meant to draw attention to this looming catastrophe but to get a consensus on how to reduce greenhouse emission. This chapter avows that there is a correlation between climate change, land and the phenomenon of the Fulani herdsmen and farmers conflicts in Plateau State and Nigeria. For about two decades now, the Nigerian state has witnessed the scourges of desertification and afforestation, especially across the Sahel region, which is eating deep into Nigeria's northern corridors (Sokoto, Zamfara, Katsina, Kebbi, Kano, Kaduna, Jigawa, Borno, Yobe, Gombe and Bauchi states, among others), such that rivers and grazing fields are drying up, farmlands are no longer as fertile as they once were, heatwaves have continued to threaten both humans and cattle, hunger and disease are on the increase, and the rising sea and water levels are causing flooding, among other impacts(Abugu & Onuba, 2015).

This phenomenon has resulted in forced migration of the Fulani, especially and their cattle in search of greener pasture across the Middle Belt and southern region of the country. The Nigerian mainstream media approach to the coverage of conflicts and climate change issues is not adequate; they often don't focus on preventing those phenomena before they occur. In keeping with the social responsibility of the media, which have the responsibility to inform as well as to educate the members of the public on issues that could fuel climate change and conflict. The media should also bring to the knowledge of the Nigerian government, the political elite as well as galvanise all social movements, through a sustained media coverage of the effects of climate change on the environment and how this is forcing, not just the Fulani herdsmen and their cattle, but also communities to migrate to other areas. These migrations have fuelled the contestation for grazing and farm lands, which have, in turn, resulted into violent conflicts in Plateau State and in Nigeria as whole. Although human rights journalism is not flawless the chapter avows that if it is fully imbibed and practiced by both journalists and their news organisations it would help in the fight against the scourge of climate change and violent conflicts in the country.

Conceptual Framework

Although the Nigerian mainstream media have a long history of covering conflicts within its borders and beyond, unfortunately, the same cannot be said of their conflict-prevention strategies. Human rights journalism (HRJ) approach as conceived by Shaw (2012), which guides this chapter contrasts sharply with other media approaches to the coverage of conflict. HRJ approach goes beyond the reportage of physical violence to looking at the underlying causative factors to prevent conflicts before they would eruption. Shaw contends that human rights journalism is a diagnostic approach of reporting which gives a critical reflection of the experiences of the victims of human rights abuses in all its ramifications. Accordingly, human rights journalism, in Shaw's opinion, is predicated on the notion that, if journalism is to be responsive and alive to its responsibilities to the society, it must focus on deconstructing the embedded structural causes of violence, such as economic and political exclusions of ethnic groups, climate change, youth unemployment, among others. In other words, human rights journalism aims to be pro-active and not just reactive to the phenomenon of conflict. It challenges journalists to be relentless as they are unyielding in their pursuit of justice for victims of human rights violations by galvanising social forces across board to bring the necessary pressure to bear on the government and policymakers to act. This model of journalism, as Shaw (2012) has noted, challenges political, economic, social and cultural imbalances that serve to undermine peoples' rights. Human rights journalism is, as he (Shaw) aptly described, as a brand of journalism with a human face. In consonance with its name, HRJ does not just investigate the actual violence but also probes indirect structural/cultural violence, such as colonial policies that accentuated and stratified ethnicity, drought, forced labour, extreme poverty, corruption, modern day slavery, famine, hate speech, racism (Ashworth, 1999; Shaw, 2012; Steiner & Alston, 2000).

Shaw (2012) explains that the opportunities inherent in HRJ enables it to go beyond just connecting people around the globe, ensures that journalists have the moral responsibility to increase awareness of the audience on their rights by monitoring, investigating and reporting all forms of human rights violations. Based on the orientation of HRJ, it can appropriately be referred to as a type of journalism with a conscience because of its humane approach to news coverage. Shaw observes that, although HRJ tries to respond to conflicts by mobilising policymakers and social movements to act, the unique advantage that HRJ has over the mainstream media perspective is its ability to be proactive and anticipatory. Through this process, it can challenge indirect structural and cultural imbalances of society as cold conflicts, instead of waiting until they manifest as direct political violence, as hot conflicts and by which time it would probably be too late. In other words, HRJ challenges rather than reinforces the status quo of the dominant voices of global and national society. As Shaw (2012, p. 41) further explained, 'Human rights-based journalism, made possible by a free and open media climate, gives early warning of conflicts and other security threats.' This position is further reinforced by Sen (1999) when he opined that India's free press, for example, was instrumental in helping the country avoid famine since independence by sensitising its policymakers to action when hunger reached life-threatening levels. If the Indian free press could help avert famine in their country, what this also suggests is that if HRJ, if it is imbibed and practiced, is effectively applied it can help push back the frontiers of climate change and thus avert conflicts occasioned by the forced migrations of cattle and humans. In the next section, attempts will be made to analyse some media approaches to the coverage of conflicts.

MEDIA APPROACHES TO THE COVERAGE OF CONFLICTS

Mainstream Media

The notion 'mainstream media' refers to dominant media, such as the radio, television, newspapers and magazines, among others. It needs to be stressed here that one of the key features of the mainstream media is that they take sides with the official rhetoric and policy position of the political and dominant elite in society. Their patterns of reporting are generally characterised by sensationalism and propaganda. Very often, the mainstream media do not undertake in-depth analysis of the underlying factors behind a conflict, but pay greater attention to the coverage of actual conflict (Carter, Thomas, & Ross, 2011). One of the critics of the mainstream media reporting, Galtung (1992) has accused them of concentrating on problems more than giving solutions, on propaganda rather than on truth, on elite rather on ordinary people.

HRJ, as stated earlier, is more concerned with how to prevent conflict, thus it would be less concerned with propaganda and sensationalism which the mainstream media are known for. Human rights journalists would focus more on the masses as victims rather than on the political elite in society.

Several examples of how the mainstream media technique to the coverage of conflict led to the intensification of conflicts and loss of lives and property abound. The mainstream media coverage of the Rwandan genocide of 1994 is one such example. Hutu extremists used Radio Television Libres des Millo Collins as propaganda tools to incite their fellow Hutus against the Tutsis (Mitchell, 2007). The aftermath of that was the genocide that shook the entire civilised world. Similarly, the Kafanchan religious conflict (between Christians and Muslims) of 1987 also highlights the alleged roles of two dominant media in northern Nigeria, that is, the Federal Radio Corporation of Nigeria, Kaduna and the New Nigerian newspaper. Hackett (2010) claims that Federal Radio Corporation of Nigeria and the New Nigerian were complicit in the conflict by exaggerating Muslim fatalities, but failed to also report reprisal attacks on Christians. This biased reporting by the two media, in the views of Hackett, led to the escalation of the conflict. Furthermore, in a Government White Paper Report of the Judicial Commission of Inquiry into the Civil Disturbances in Jos of September 2001, the Miyetti-Allah Cattle Breeders Association of Nigeria accused the General Manager of the state's-owned Plateau Radio and Television Corporation (PRTVC), Mr Joseph Ari for allegedly being:

[...] against Islam, which we have earlier observed is equated with ***Hausa/Fulani*** *[...] (my own emphasis). Despite being a public property, funded by money that accrued from both Muslims and Christians' (Plateau State Government, 2009, p. 190).*

(Although Hausa and Fulani are two distinct ethnic groups, they are, however, the two ethnic groups sometimes in conflict with the Plateau State's ethnic communities. The concept is used here to denote them as they also used it to refer to themselves in the quote above). Again, the Plateau State's indigenous ethnic communities have also accused the Hausa services of the British Broadcasting Corporation (BBC), Voice of America (VOA), Radio France International and especially the Daily Trust newspaper of producing a daily propaganda against the state and its people since the outbreak of the conflict in September 2001(Plateau Indigenous Development Association Network, 2010). Taken together, these few examples of the mainstream media reporting of conflicts are fraught with allegations and counter allegations characterised by propaganda and biases aggravated the conflict.

Human Rights Reporting

It must be mentioned here that human rights reporting is an important notion in global journalism practice. Although human rights reporting is somewhat like human rights journalism, its scope and capacity is severely limited compared to HRJ. Reporting human rights approach is directed towards reporting all kinds of human rights violations and facilitating the process of bringing conflict parties to the negotiating table for the purposes of dialogue. As stated earlier, this approach to reporting conflict lacks the resources to undertake proper analyses of the triggers of conflict and their structures so to prevent any violent episode of conflict. Human rights reporting's interest is on reporting issues of rights violations that require media coverage. One of the advantages of HRJ is that it sustains pressure on human rights issues through its media coverages until victims receive justice and perpetrators are brought to justice. For example, through HRJ reports have been tendered as evidence against alleged war crime suspects in countries, such as Yugoslavia, Kenya, Rwanda, and Sierra Leone, among others (Shaw, 2012). Thus, human rights reporting is not robust enough to prevent conflicts before they erupt because it focuses more on actual conflicts.

Peace Journalism

Peace journalism which has been variously designated as 'conflict sensitive' or 'solution journalism' is both conceptually and practically developed to equip and change, not just the attitudes and orientations of journalists (reporters, editors, news media professionals, and media owners), but also the attitudes of the members of the public. This is so that they would become conscientious and responsible in their reportage of conflicts and to contribute to the process of making and keeping peace (Hanitzsch, 2007; Lynch & MacGoldrick, 2005; Shinar, 2003). Based on this definition, peace journalism is meant to assist, especially news media professionals to make informed decisions about, not just how stories are to be published, but crucially, how they are to be framed. This is because news framing, as Danaan (2017) has avowed, is central to any journalistic discourse. Hence, news framing is not a product of chance, rather, it is a product of a deliberate and conscious judgement (Frame, 2013). Peace journalism, based on its conception, seeks to assist society at large with opportunities to consider as well as to value nonviolent responses to conflict (Shaw, 2012). Peace journalism, in the opinion of Lynch (2008), takes the approach of advocacy, which is different from the approach of the mainstream media whose focus is intended toward the reportage of the physical conflict.

All these media approaches to reporting conflict described how very often journalists and content creators frame their news reports. The reality is that a lot goes into news construction and production, such as external influences, ownerships and control, organisational pressures, not to mention the journalist's own ideologies (Dennis & John, 1984). News media in general are fraught with these challenges. Although HRJ is not perfect, the point of departure for the human rights journalism is that its journalists are specially trained to proactively prevent violent conflict by highlighting those issues that are likely to precipitate conflict. It therefore follows that it is not every journalist that can possibly be a human rights journalist, because the mainstream media journalists often are not able to rise above their ethnic, religious or political affiliations. Besides, the mainstream mass media, within the context of the propaganda model of Herman and Chomsky (1998) always seek to serve the interests of the political elite and corporate bodies. As it has been noted earlier, the mainstream media are preoccupied with reporting facts, truth and people, peace journalism, on the other hand, is concerned with conflict resolution and the promotion of peace initiatives (Sule, 2015). Despite the great opportunities peace journalism offers towards peace initiatives and conflict resolution, it lacks the sophistication to analyse possible triggers of conflict and so to proactively prevent them before they erupt. HRJ, on the other hand, based on its ethical commitment has a responsibility and the resources to not only witness but crucially reports on all forms of human rights violations (Shaw, 2012). HRJ can help towards the prevention of climate change since, as Sen (1999) claimed, India's free press was significant in helping the country avoid famine since they gained independence by sensitising its policymakers to action when hunger reached life-threatening heights. Thus, if human rights orientated journalism is practiced by both journalists and their news corporations it assist in preventing the outbreaks of violent conflicts in Plateau State and in Nigeria.

Conflicts in Nigeria: A Background

The Nigerian state is the creation of the British colonial administration through the final amalgamation of 1914(Mangvwat, 2013). Thus, this process, brought together different communities without due consideration for their ethnicities, religions, cultures, traditions, polities, histories, among other differences. By forcefully bringing together these multicultural groups and communities, it created what has now become known as ethnic 'majorities and minorities.' Although, the ethnic majorities and minorities issue is not peculiar to Nigeria because it is one of the oldest problems that has persisted around the world, whether it is the Palestinian-

Israeli, Turkish-Kurds, Arab versus Black Africans in the Sudan, the Berbers and the Saharawi in Morocco, among others (Dzeka, 2014), it is the root cause of ethnic conflicts in Nigeria. Several scholars (Gofwen, 2004; Mangvwat, 2013; Onwubiko, 1985) have argued that before the British conquest and subsequent establishment of the colony and protectorates of northern and southern Nigeria, hardly did this problem exist. This is because most of the ethnic groups were independent polities, autonomous communities, chiefdoms, sovereign kingdoms, city-states and village confederations(Mangvwat, 2013).Others have posited that the colonialists used ethnicity as a core strategy in their administration of the country, which today is believed to be at the heart of the myriads of conflicts(Attoh & Soyombo, 2011; Salawu & Hassan, 2011).Chabal and Daloz (1999)have strongly argued that the leftover of 'the structure of the post-colonial state has enabled those African power elite who held power to instrumentalize ethnicity into political tribalism.'Samuel (2012)has outlined some of the country's ethnic conflicts that involved ethnic communities to include: the Ijaws and Urhobo, on the one hand, and Itsekiris, on the other, in Warri; Aguleri and Umuleri in Anambra; the Ife/Modakeke, Yoruba (Sagamu) and Hausa (Kano) in the West; Jukuns/Tivs in the Middle Belt area of the country. What has fuelled conflicts in Nigeria whether it is ethnic, religious or political, is mainly the unresolved issue of citizenship (Alubo, 2006). It is unresolved problem that has created the 'indigenes/settlers' conundrum. This, therefore, provides the background and the context upon which conflicts in Nigeria can be better understood. The next section discusses to the role of the media in the coverage of conflicts in Nigeria.

Coverage of Conflicts in Nigeria: The Role of the Media

The Nigerian state is composed of over 370 ethnic groups (Alubo, 2006) and based on this composition, it was to be expected that the propensity for conflict would be high. The British colonial policies, especial the final amalgamation of 1914, which brought the diverse ethnic communities together, accentuated differences and bred misunderstandings, that later led to conflicts, which were sometimes violent (Salawu, 2009). The country, has had more than its fair share of conflicts, especially since the inauguration of the 4th Republic in 1999(Galadima, 2010). The country's political landscape has been replete with a myriad of conflicts with attention-catching-headlines in some of the print media thus: 'Sharia Tears North Apart' (Tell magazine, March, 2000); 'Bloodbath in Kano and Taraba' (Newswatch magazine, October 29, 2001); '13 killed in fresh Jos violence' (Thisday newspaper, January 2, 2002); 'Wari: 7 Die in fresh violence' (Thisday newspaper, January 26, 2004); 'Militia men kill 52, burn 300 in Benue, Delta' (The Guardian, May 19, 2004)(Alubo, 2006, pp. 1-2).

Although the colouration and complexion of these conflicts bear the hallmarks of ethnicity and religion, both variables are not the driving force of conflicts in Nigeria, but levers used by the country's political elite to rally support in pursuit of what Aapengnuo (2010, p. 1) would refer to as 'power, wealth and resources.' Since the variables of ethnicity and religion are so strong and fluid, the Nigerian media system has not only become enmeshed in it, but, indeed, as Akinfeleye (2003, cited in Okidu, 2011) argued, set the ethnic and/or sectional tone, on the eve of independence in 1960 and after independence. Scholars, such as Oso (2012, p. 278) reasoned that 'the emergence of political parties and the increasing prospect of independence altered the political orientation of the Nigerian press.' From this point on, the media shifted their concentration from the colonial government to organs of political parties and leading politicians of the dominant political parties that were regionally and ethnically-based (The West African Pilot of Dr Nnamdi Azikiwe became the mouthpiece for the eastern region, The Nigerian Tribune for the West and The New Nigerian for the North).Galadima and Soola (2012) have chronicled newspapers and magazines that have proliferated since after independence and how they have continued the promotion of regional ethnic/political interests thus: the Daily/Weekly Trust is pro Hausa/Fulani of the core north; Newsmonth, Newsgate, Pinnacle, Prism, Ultimate, Citywatch, Prime, News line, Africa news, The Nigeria Standard and Vision for the Middle Belt; The Punch, The Western Voice, the Westerner, Odua News, Akede Oduduwa, for the West; Daily Champion, and Daily Sun for the south-east group during the last 2007 general elections in Nigeria. The 30-month brutal Nigeria Civil War which lasted from 1967-1970 had as its root cause, ethnic rivalry (Samuel, 2012). Abati (2000) further posited that the origin of ethnicity in the Nigerian press is traceable to the entanglement between journalism and politics in the early days of Nigeria's independence. Since then the media coverage of issues in general, especially conflicts in the country are always mirrored from the perspectives of ethnicity and religion (Ishaku, 2012; Nnoli, 1978; Plateau Indigenous Development Association Network, 2010; Yusha'u, 2015). To illustrate how the media coverage of conflicts in Nigeria takes the ethnic tone, the next section examines the media coverage of the Plateau State's indigenous communities vs Hausa/Fulani group.

Media Coverage of the Plateau State's Indigenous Ethnic Communities vs. Hausa/Fulani Group Conflict

A British colonial official by the name Colonel H. W. Laws led the colonial forces that occupied the Plateau area in 1902(Mangvwat, 2013; Plateau State Government, 2004). From this point on, the Plateau area was annexed to Bauchi Division by the colonial occupying forces. Plateau State is positioned in the central Nigerian area known as the Middle Belt which is home to over 370 ethnic groups (Alubo, 2006). A combination of factors ranging from British colonial policy of racial segregation that divided groups based on their ethnicities, economic and political exclusion to poor leadership led to the feeling of frustrations that would later snowballed into violent conflicts, especially between the Plateau State's indigenous ethnic groups and the Hausa/Fulani(International Crisis Group, 2012; Mangvwat, 2013).For example, within the Jos area of the Plateau State capital, the city was divided into three main clusters for different ethnic communities: the first category was the Government Residential Area (GRA), where Europeans and other foreigners, and the cream of colonial elite occupied; second, the Native Town, was inhabited by the Hausa ethnic community; third, there was the Township, which was composed of other Nigerians and foreigners(Best, 2007). This colonial structure emphasised ethnic differences which exists in Plateau State even today, although not in its strict form. But up to now, it characterises inter-ethnic relations in state. In almost all the seventeen local government areas of the state, Hausa, Igbo, Yoruba, natives and foreigners live in their separate quarters or districts (Ochonu, 2008). It is the same British colonial policies of ethnic segregation of north and south that was a major cause of the first Sudanese civil war of 1955-1972 (Mazrui, 2000). Scholars, such as Egwu (2004); Kalejauye and Alliyu (2013); Nnoli (2008) have blamed this colonial policy of segregating (balkanizing) the various ethnic communities.

The Plateau State conflict has been variously discussed by scholars from a variety of perspectives: ethno-religious, economic, political, media assessment, among others (Galadima, 2010; Gambo & Dakop, 2010; Gofwen & Ishaku, 2006; Higazi, 2011; Idris, 2010; Ishaku, 2012; Jega, 2003; Kaigama, 2012; Katu, 2016; Lohor, 2012; Mangvwat, 2013). The conflict has spanned nearly twenty years. Studies on the conflict especially in the last two decades have attributed it to the contestation for land, economic and political access in its capital Jos which later spiralled over to other parts (Ishaku, 2012; Kaigama, 2012; Mangvwat, 2013). The Plateau State Indigenous Development Association Network (an umbrella body of the state's indigenous ethnic communities) claimed that the history of the conflict, in recent times, goes back to the creation of Jos north local government area by the General Ibrahim Babangida administration in 1991. The network argues that the creation of the local government was meant to cede Jos, the political capital of the state and its economic nerve centre to the Hausa/Fulani. Other events that further exacerbated the conflict was the appointment of Alhaji Sanu Mato as caretaker committee chairman of the local government to the displeasure of the indigenous ethnic groups, especially Afizere, Anaguta and Berom triad who are thought to be the owners of Jos (Ishaku, 2012). Another around of conflict erupted on the 28 November 2008 during the Jos north local government elections. Trouble started when supporters of the two main political parties Peoples' Democratic Party (PDP) and the All Nigerians Progressive Party (ANPP) clashed before the final results were announced. The ANPP candidate, Aminu Baba, fearing that the result of the elections was not going in his favour, raised alarm that the elections were being rigged in favour of the PDP candidate, Timothy Buba (Galadima, 2010). Buba, the candidate of the PDP was the candidate of the Plateau State indigenous communities, while Baba was the candidate of the ANPP and the Hausa/Fulani. The conflict spread to other local governments of the state like Jos south, East, Barkin Ladi, Bassa, Bokkos and Riyom.

In a report by Bulus Gambo in The Sunday Standard of March 7, 2010 entitled 'Act now to avert Jol crisis escalation,' (Gambo, 2010), the reporter claimed that during the conflict, Fulani herdsmen with their families and cattle numbering over 5, 000 from the northern part of the county streamed into Riyom local government area of Plateau State. The reporter further claimed that the Fulani descended on the indigenous Berom communities in the area, killed some, displaced others and destroyed their farmlands. This report by Gambo resonates withChinwoku (2017, p. 30) position that 'the Fulani herdsmen carry sophisticated firearms and attack their target communities [...].'This scenario appears to fit into the popular narrative of forced migration of the Fulani herdsmen because of the drought that has eaten into the northern part of the country from the Sahel region (Abugu & Onuba, 2015). In the view of Boko (2007, cited in Abugu & Onuba, 2015), to escape from the vagaries of the drought-stricken areas of the north, the Fulani and their cattle migrate to the southern parts of the country through the Middle Belt area (Plateau, Nasarawa, Kogi & Benue States), their gateways. Often, in the process of this migrations by the Fulani and their cattle, due to poor rain fall and grazing land caused by climate change, communities are sacked and their lands taken over (Abugu & Onuba, 2015).

The Plateau State conflict must be placed within its historical context of how the indigenous ethnic groups, especially the Anaguta and Berom of the Jos area were disposed of their land resource. Historical accounts of the development of the Plateau area and especially Jos was because of some events: the discovery of tin in large commercial quantities, the construction of railway from Kaduna junction to Jos in 1927 and agricultural opportunities (Mangvwat, 2013; Plateau Indigenous Development Association Network, 2010). These events, especially 'tin rush', in the opinion of Mangvwat, between 1904 and 1914 increased the tempo for the prospecting for tin, hence serious scramble for land, an important resource of the Berom and Anaguta whose livelihood revolved around it (p, 99). With the influx of over 50 foreign companies in 1911; 70 in 1912; 72 in 1926; and 83 in 1928, Mangvwat argued that these further put pressure on the need for more land which due to the mountainous terrain was a scarce resource compared to other parts of the country such as Kaduna and Bauchi divisions. This scramble for the partition of land led to the Anaguta and Berom ethnic communities being forcefully ejected from their ancestry lands and their crops destroyed without compensation. It had to take the Berom uprising of 1941 on the land issue which was organised and widespread before compensation began to be paid. Mangvwat further avowed that this situation was replete across Berom and Anaguta villages of Gyel, Zawan, Shen, Vwang, Du, Riyom, Vom, Gashish, Ropp, Ganawuri, Bassa, Kwal and Jere districts.

It is worth noting that prior to the establishment of the Pax Britannica in northern Nigeria,Plotnicov (1972) contends that, there had been successful raids by the Hausa/Fulani in obtaining slaves on the Jos Plateau. However, he argues that they were not successful in obtaining lands in the area to establish political control. The belief of the Hausa/Fulani that it is their ordained rights to rule over non-Hausa/ Fulani ethnic groups was summed up thus: 'But now it seemed that what could not be achieved by force of arms was granted by the grace of God via British colonial intervention' on the Plateau (Plotnicov, 1972). Similarly, Plotnicov summed the indigenous communities' frustrations as follows: 'Their land had been overrun by their former enemies-the Hausa/Fulani by means of superior might of the Europeans [...]' (p, 8). Thus, this provides the context for a better understanding of earlier discussion on recent conflicts that ensued between the Plateau State indigenous ethnic communities and the Hausa/Fulani ethnic group.

The conflicts have been widely covered by both local and international media (Abugu & Onuba, 2015; Ishaku, 2012; Plateau Indigenous Development Association Network, 2010). The media were not only accused of bias coverage by both the Plateau State indigenous communities and the Hausa/Fulani ethnic group, but it also brought to the fore the ethnic influence on the media coverage. Galadima (2010) contends that the media coverage of the conflict was fraught with allegations of bias, especially by the ethnic groups involved in the conflict (Plateau State indigenous communities and the Hausa/Fulani). These accusations were championed by leaders of the various ethnic communities. Galadima vows that the Plateau State Government and the state's indigenous political elite representing its local and national assemblies, unequivocally condemned the reports of the conflict by the Hausa services of Voice of America (VOA), British Broadcasting Corporation (BBC), and Radio Deusche Velle (*Thisday* 29, September 2001, p. 8).On their part, the Hausa community under the aegis of Jasawa (the umbrella body of the Hausa Development Association in Plateau State) accused the Plateau Radio and Television and the state-owned newspaper, The Nigeria Standard of bias and sectionalism (Galadima, 2010). Musa and Ferguson (2013), in their study of the Plateau State conflict had observed how some newspapers used frames that were aimed at polarising Plateau State thus 'Us' and 'Them.' They further reinforced their point by citing some stereotypical frames used by two newspapers they studied: the Daily Trust and ThisDay. For example, the Daily Trust used such metaphors and stereotypes like 'marauders' and 'Dear sons of the soil' in their headline news to refer to the Plateau State's ethnic communities.

ThisDay newspaper used demeaning adjective such as 'mercenaries' to refer to the Hausa/Fulani ethnic community. Usage of such derogatory lexical were aimed at demonising the other ethnic group and further intensify the conflict.

The media coverage of the Plateau State conflict has been marred by a plethora of accusations levelled against the media for their slanted and biased coverage (Galadima, 2010; Ishaku, 2012; Musa & Ferguson, 2013). The media, rather than be on the vanguard of bringing an end to one of the protracted conflicts Nigeria has witnessed in recent times, the media became enmeshed in producing frames that were aimed at advancing the cause of their ethnic communities. Condemnations of the media's patterns of coverage of the Plateau State conflict are well documented. For example, Rasaq (2012) points to a report by the Daily Independent of 22 February 2010 reported thus, 'It is obvious that certain sections of the media have taken sides in reporting of events in the recurring conflict in Plateau State, and this has the capacity to cause more harm than good.' Similarly, The Vanguard of Saturday, 27 March 2010 quoted the General Officer Commanding 3 Division, Rukuba, Major General Saleh Maina thus, 'I was very disappointed that some parts of the print media misconstrued, misunderstood, misrepresented and misinformed the public and was apportioning blame on the military.' In his scathing remarks on the way journalists framed the Plateau State conflict, Danaan (2017, p. 52) posited that 'journalists use of the media to achieve goals set by either ethnic or religious communities to which they have pledged their allegiance' rather than the ethic of journalism was disappointing. These patterns of media coverage of conflict which ignores their media ethics but are motivated by their ethnic and/or religious logic shows the failure of the kind of critical analysis and investigative depth required in journalism practice (Okidu, 2011, cited in Danaan, 2017). If the media, especially the Nigerian newspapers are to redeem their already tainted image before the Nigerian public and the world, they must, as Akinfeleye (2003) warned, respect and chart a new course of action by setting a new agenda for adherence to the social responsibility norms of the press, a core value of journalism practice.

MEDIA COVERAGE OF CLIMATE CHANGE IN NIGERIA

General Outlook

One of the hot button issues of the contemporary times, among a litany of others, is climate change. This multi-faceted policy issue is dominating discourse in the global community because of its devastating impact, which is already being felt across the nooks and crannies of the world. The United Nations-sponsored conference on this phenomenon in Paris in 2015 has taken the discourse to a new height. The world today is witnessing increased human activities on the planet which are causing environmental and social degradation (Deben, 2015). Although the impact of climate change is global, the way it affects continents and countries vary considerably. Experts in the field, such as Marino and Ribot (2012) and Constable (2017) have noted that, ironically, the distribution of climate change impacts is already disproportionately affecting those countries who have contributed the least compared to the industrialised countries of the world such as India, China, Russia and America to historical emissions of greenhouse gases (GHG). Enahoro (2012) contends that recent studies on climate change effects are greater in Africa than in the rest of the world. The reason for this, as Enahoro avowed is because the continent is 0.50C warmer than it was some 100 years ago. On the global scale and based on a study by World Bank on greenhouse emissions, Nigeria has been ranked 35th out of 150-member states (www.worldbank.org). This report does not in any way make for pleasant news because of the ever-increasing greenhouse emission in Nigeria.

Nigerian Government and the Fight Against Climate Change

The federal government of Nigeria, through its Ministry of Environment and appear committed to combating this unfolding catastrophe. The country has been actively engaged in international climate change negotiations since it became a signatory and a party to the UN Framework Convention on Climate Change (FCCC) in 1994 ratifying its Kyoto Protocol in 2004(Environment, 2015). It is on record that Nigeria submitted its First National Communication (FNC) in 2003 and a second National Communication in February 2014). Nigeria has also hosted several Clean Development Mechanism projects, and other projects supported by the Adaptation Fund. Other efforts by the Nigerian government in this regard is the approval of the Nigeria Climate Change Policy Response and Strategy in September 2012 by the Federal Executive Council and Nigeria's participation at the UN sponsored conference in Paris in 2015. Nigeria's active role in the fight against climate change and other environmental degradations is understandable because the country is hugely impacted by the effects of climate change. Enahoro (2012, p. 460) posits that:

The changing weather patterns are already creating new complex emergencies where poorer countries are continually affected by drought and floods, often accompanied by outbreaks of infectious diseases such as cholera and Rift Valley Fever, which were thought to have been eradicated, have now reappeared making many communities live by almost permanent disaster conditions.

Even within the same country, the northern part, has in recent times been vulnerable to the vagaries of both drought and flooding. Communities in states like Benue, Kebbi, Sokoto, Nasarawa, Gombe, Katsina, among others, have been submerged due to flooding in 2017; houses and agricultural produces have been destroyed. It is, therefore, not surprising when Pope Francis warned in his message thus: 'Climate change [...] represents one of the principal challenges facing humanity in our day. Its worst impact will probably be felt in developing countries [...]' (Deben, 2015, p. 16).

Climate Change, Force Migration and Conflict

Experts and scholars have been unequivocal in their predictions and forecast of the dire consequences of climate change on the global community but more on the developing nations of the world (Constable, 2017; Deben, 2015; Enahoro, 2012; Marino & Ribot, 2012). The federal government of Nigeria has, no doubt, taken a leading role in its policy response to the menace and challenges posed by climate change. Nigeria has had a long history of violent conflicts (Alubo, 2006; Danfulani, 2006; Fwatshak, 2005-2006; Gofwen, 2004). The media have been preoccupied with the coverage of these conflicts in Plateau State and other conflict areas such as the fight against Boko Haram in the north-east of the country. Some of these conflicts are linked to the changing demography of the country because of the changing weather patterns (Abugu & Onuba, 2015; Enahoro, 2012). Among the negative effects of climate change includes: the depletion of the ozone layer, which is the major cause of heat wave, dryness of the rivers and lack of pastures for cattle to graze, lack of adequate rainfall. Indeed, experts like Leed (2006), have warned that the unattended negative effects of climate change will lead to increased environmental degradation and will also lead to forced migration of communities in their millions from the Sahel region to other regions of the country especially the Middle Belt and the southern parts. This will in turn, as earlier noted, lead to increased situations of conflicts, especially between the Fulani herdsmen and farming communities along the Middle Belt and the southern parts of the country. Thus, the federal government's report has highlighted some of these dangers that climate change portends for Nigeria and Nigerians (Environment, 2015). The government outlined its strategic goal of the Climate Change Policy Response and Strategy objectives thus:

- Implement mitigation measures that will promote low carbon as well as sustainable and high economic growth;
- Enhance national capacity to adapt to climate change;

- Raise climate change related science, technology and P & D to a new level that will enable the country to better participate in international scientific and technological cooperation on climate change;
- Significantly increase public awareness and involve private sector participation in addressing the challenges of climate change;
- Strengthen national institutions and mechanisms (policy, legislative and economic) to establish a suitable and functional framework for climate change governance.

Furthermore, in the federal government of Nigeria's determination to achieve its goal in combating the incident of climate change, it took firm steps towards this objective to reduce vulnerability; minimise risks by improving adaptive capacity by setting up a thirteen sector-specific strategies, policies, programmes and measures that are actionable by the Federal, State and local governments, civil society, private sector, communities and individual thus:

- Improve awareness and preparedness for climate change impacts;
- Mobilise communities for climate change adaptation actions;
- Reduce the impacts of climate change on key sectors and vulnerable communities;
- Integrate climate change adaptation into national, sectorial, State and Local Government planning and into the plans of Universities, research and educational organisations, civil society organisations, the private sector and the media (Environment, 2015, pp. v-vi).

While the Nigerian government's policy/initiatives, as outlined above, look plausible and quite capable of turning the tides of this crisis that is already upon humanity, the Nigerian media that are supposed to, not only help step down this policy/initiative for the audiences but through their investigative journalism help to keep the climate change issue on the front burner of national discourse are not leading from the front. In a study by Udoaka (2015) on Nigerian newspapers coverage of environmental issues to mark World Environmental Day, the result was disappointing. Udoakah sampled four leading Nigerian newspapers: The Vanguard, The Punch, Guardian and Thisday. In his research, which covered the period 1-14 June 2004 and 1-14 June 2005, 60 per cent of environmental content, Udoakah claimed, was published in the inside pages of the newspapers and 40 per cent was published on the back pages of the papers. What this shows is that environmental issues were not given the prominence they deserved. This pattern of media coverage is consistent with the mainstream media, human rights reporting and peace journalism. They tend to emphasize physical issues such as the coverage of conflict, ignoring the underlying causative factors of the phenomenon. The media have a social responsibility to keep the society informed and to educate the people on issues such as climate change so that they too could close ranks with the government in the fight against the threats that they pose to both humans and the environment. With the right information from the media, citizens would be better informed about the danger of indiscriminate felling of trees, carbon monoxide, dumping of refuse in drainages, building on water channels, among others. Two television stations in Nigeria, namely, Television Continental (TVC) and Channels Television have been on the forefront of efforts in this regard through their daily features on climate issues and the environment.

Towards a Human Rights Orientated Journalism Approach to Coverage of Conflicts and Climate Change in Nigeria

The issue of farmers and Fulani herdsmen conflicts over grazing land has been a recurring theme in Plateau State and in Nigeria at large. Plateau State located in north-central Nigeria sits at the crossroads between the north and the south of the country, which is characterised by low rainfall and drought-stricken conditions (Shukla & Futrell, 1993). Desertification and drought are eating deeper and deeper into the northern region of Nigeria from the Sahel region, which has necessitated the southward movement of the Fulani herdsmen every year with their cattle for greener pasture as well as to escape from tsetse flies (Abugu & Onuba, 2015; Genyi, 2014). In a study by Badejo (1998), he contended that the savannah is creeping fast into the rainforest in Nigeria due largely to human factors, such as bush-burning, agricultural expansion/intensification, fuel-wood extraction, and climatic/physical factors such as rainfall variability (Macaulay, 2014). Therefore, the fierce contestation for access and utilisation of resources such as land, water and pasture in the central Nigerian areas; especially Plateau, Nasarawa and Benue States between the indigenous ethnic communities and the Fulani herdsmen takes place within the context of the peasant and nomadic economic production systems adopted by both ethnic groups. Indeed, as Genyi (2014) has observed, the Fulani system of production revolves around cattle rearing. He maintained that of all the ethnic groups in Nigeria, the Fulani herdsmen remain one of the few ethnic groups that have resisted efforts to modernise grazing as supported by Barkindo, Makanjuola, and Martins-Aginam (2017) and because of their system of production and consumption they have remained largely unaltered in the face of population growth and modernisation.

CONCLUSION

This chapter has surveyed the media coverage of conflicts and climate change issues in Plateau State, Nigeria. It examined other media reporting techniques to in respect of conflicts and climate change, such as the mainstream media, human rights reporting, peace journalism and human rights journalism. Having analysed these approaches, although their potential for good was noted, their shortcomings were also very apparent. There has been so much media coverage of physical conflicts in general but so little of climate change issues in Nigeria. The chapter evidenced the strides the Nigerian government, through its Ministry of Environment is doing in developing policy-issues and connecting with relevant international organisations on those climate change issues. However, there has not been a concerted media effort in this regard. It is also clear from this discourse that there is a nexus between land, conflicts and climate change problems as shown in Plateau State and in the rest of Nigeria. Human rights journalism approach, based on its orientation and proactive strategy to the reporting of human rights issues is well-suited for the coverage of climate change in Nigeria. One of the ways HRJ can collaborate with the Nigerian government is by helping to frame government's policy, to ensure timely and clear reporting of those issues to the general public. It can equally play a vital role in the dissemination of government policies/initiatives in the different vernacular languages in the country. Thus, this could help resolve the struggle for land caused by climate change that could result into conflict in Plateau State and in the whole of Nigeria.

SOLUTIONS AND RECOMMENDATIONS

In the light of the above discussion, the chapter makes the following recommendations towards the fight against climate change and stem the tide of conflicts arising from it:

- Human rights journalism it is a conflict-prevention model that goes beyond the coverage of physical conflict and addresses the deep causes of the problems. Based on this direction of human rights journalism it can help alert Nigerian policymakers and civil society organisations on the dangers of drought, climate change, global warming, among other looming catastrophes that may lead to violent conflicts.

- The Nigerian Ministry of Environment is doing a lot both at the national and the international level on the important issue of climate change, but most Nigerians are not aware of this danger posed by global warming. Human rights journalism can in various ways (through editorial analysis, commentaries, reports, including vernacular languages) to alert and educate the members of the public.
- Since human rights journalism is in the forefront of preventing human rights violations, it could be very useful in educating the Fulani herdsmen to embrace new methods of cattle rearing such as cattle ranches and grazing reserves as it is being done in most countries of the world today. This would be of greater benefits to them and their cattle as it would provide access to veterinary inspectors, water, feeds, among others. This, human rights journalism can help dissiminate information to herdsmen via documentries, drama, and so on, in their own language.
- Human rights journalism can facilitate workshops, seminars and conferences, especially to the mainstream media reporters, editors and other news media professionals in conflict-prevention strategies to be guided by human rights considerations. Since news framing by journalists is a conscious, deliberate and a well-thought out decision, through careful selection of frames they can set agenda for peace.

REFERENES

Aapengnuo, C. M. (2010). Misinterpreting ethnic conflicts in Africa, *Joint Force Quarterly, 58*(4), 13-17.

Abati, R. (2000). *The press, politics and society in Nigeria*. Lagos: Tosen Consult.

Abugu, S. O., & Onuba, C. O. (2015). Climate change and pastoral conflicts in the Middle Belt and south-east Nigeria: Implication on human resource of the regions. *Global Journal of Human Resource Management*, *3*(5), 44–51.

Akinfeleye, R. A. (2003). *Fourth estate of the realm or fouth estate of the wreck: Imperative of social responsibility of the press (inaugural lecture series)*. Lagos: University of Lagos Press.

Alubo, S. O. (2006). *Ethnic conflicts and citizenship crises in the central region.* Ibadan: University of Ibadan.

Ashworth, G. (1999). *The silence of women.* Cambridge, UK: CUP.

Attoh, F., & Soyombo, O. (2011). The politics of ethnic balancing in Nigeria. *International Journal of Sociology and Anthropology, 3*(2), 40–44.

Badejo, M. (1998). Agroecological restoration of savanna ecosystems. *Ecological Engineering, 10*(2), 209–219. doi:10.1016/S0925-8574(98)00012-3

Barkindo, A., Makanjuola, H., & Martins-Aginam, A. (2017). *2017 research report from the valley of death: Memory, healing and inter-group dialogue in southern Kaduna.* Abuja: Abbey Digital Prints.

Best, S. G. (2007). *Conflict and peace building in Plateau State, Nigeria.* Ibadan: Spectrum.

Carter, D. L., Thomas, R. J., & Ross, S. D. (2011). You are not a friend: Media conflict in times of peace. *Journalism Studies, 12*(4), 456–473. doi:10.1080/1461670X.2010.530972

Chabal, P., & Daloz, J.-P. (1999). *Africa works: Disorder as political instrument.* Oxford, UK: International African Institute.

Chinwoku, E. C. (2017). Glamorisation of Fulani herdsmen's criminality and its implications on national security in Nigeria. *Online Journal of Arts and Social Sciences,* 2(2), 29–51.

Constable, A. L. (2017). Climate change and migration in the Pacific: Options for Tuvalu and the Marshall Islands. *Regional Environmental Change, 17*(4), 1029–1038. doi:10.100710113-016-1004-5

Danaan, G. N. (2017). A theoretical discourse on the mediatisation and framing of Jos ethno-religious conflicts. *Journal of Arts & Humanities, 6*(2), 43–56. doi:10.18533/journal.v6i2.953

Danfulani, U. H. D. (2006). *The Jos peace conference and the indigene/settler question in Nigerian politics.* Paper presented at the Jos Peace Conference, Jos.

Deben, L. (2015, June 26). The Pope's critics are acting like children: The environmental encyclical makes us squirm because it expose our greed and recklessness with the powerful searching for the Gospel. *Catholic Herald,* 16-17.

Dennis, E., & John, C. M. (1984). *Basic issues in mass communication: A debate*. New York: Macmillan.

Dzeka, T. T. (2014). Ethnicity as a factor in Nigerian disunity. In V. Egwemi, T. Wuam & C. S. Orngu (Eds.), Federalism, Politics and Minorities in Nigeria: Essays in Honour of Professor G. N. Hembe (pp. 448–451). Lagos: Bahiti & Dalila Publishers.

Egwu, S. G. (2004). *Ethnicity and citizenship in urban Nigeria: The Jos case 1960-2000* (PhD Thesis). University of Jos, Jos.

Enahoro, A. B. (2012). *Mitigating ICT effects on climate change in Nigeria*. Uyo: BSM Resources Limited.

Environment, F. M. o. (2015). *Nigeria's Intended Nationally Determined Contribution: Conference of Parties to the United Nations Framework Convention on Climate Change (COP-UNFCCC)*. Abuja: Federal Ministry of Environment.

Frame, J. D. (2013). *Framing decision: Decision making that accounts for irrationality, people, and constraints*. San Francisco: Jossey-Bass.

Fwatshak, S. U. (2005-2006). Reconstructing the origins of the peoples of Plateau State: Questioning the 'we are all settlers' theory'. *Journal of historical society of Nigeria, 1*(16), 122-140.

Galadima, J. D. (2010). *Brothers against brothers: The press, identity politics and conflict management in northern Nigeria*. Tripatide Services Limited.

Galadima, J. D., & Soola, E. O. (2012). The press and group identity politics in Nigeria: A study of newspaper framing of the 2003 and 2007 presidential elections. *Journal of Communication, 10*(1), 1–27.

Galtung, J. (1992). *Global glasnost: Towards a new world information and communication order?* Cresskill, NJ: Hampton Press.

Gambo, B. (2010, March 7). Act now to avert Jol crisis escalations. *Sunday Standard*.

Gambo, B., & Dakop, K. (2010, January 31). Who wants Plateau on its knees? *The Sunday Standard*.

Genyi, G. A. (2014). *Ethnic and religious identities in shaping contestation for land based resources: The Tiv farmers and pastoralists conflicts in central Nigeria*. Paper presented at the Annual Conference on Ethnic and Religious Conflic Resolution and Peacebuilding, New York, NY.

Gofwen, R. I. (2004). *Religious conflicts in northern Nigeria and nation buidling: The throes of two decades 1980-2000.* Kaduna: Human Rights Monitor.

Gofwen, R. I., & Ishaku, J. (2006). *State of emergency: Our stand by Plateau patriots.* Lagos: Bookhouse.

Hackett, R. I. J. (2010). Managing or manipulating religious conflict in the Nigerian media. *Journal of Sustainable Development in Africa, 12*(3), 1–28.

Hanitzsch, T. (2007). Situating peace journalism in journalism studies: A critical appraisal. *Conflict and Communication, 6*(2), 1–9.

Herman, E., & Chomsky, N. (1998). *Manufacturing consent: The political economy of the mass media.* New York: Pantheon Books.

Higazi, A. (2011). *The Jos crisis: A recurrent Nigerian tragedy.* Bonn: Friedrich-Ebert-Stiftung.

Idris, H. (2010, April 20). Jos crises threat to national security-Miyetti Allah News. *Daily Trust.*

International Crisis Group. (2012). *Nigeria's indigene-settler divide.* London: Routledge.

Ishaku, J. (2012). *The road to Mogadishu: How Jihadist terrorism tears Nigeria apart.* Hamptul Press Ltd.

Jega, A. (2003). *Identity transformation and identity politics under structural adjustment in Nigeria.* Kano: Afrikainstitute/Apsalla/CRD.

Kaigama, I. A. (2012). *Peace, not war: A decade of interventions in the Plateau State crises (2001-2011).* Hamtul Press Ltd.

Kalejauye, P. O., & Alliyu, N. (2013). Ethnic politics and social conflicts: Factors in Nigeria's underdevelopment. *Journal of International Social Research, 6*(27), 251–262.

Katu, N. N. (2016). *Media, Conflict Audiencess and the Dynamics of Information Dissemination in Plateau State, Nigeria: Is the Tail Wagging the Dog?* (PhD Thesis). Ohio University, Athens, OH.

Leed, K. T. (2006). *Conflict resolution theory and practice: Integration and application.* Manchester, UK: Manchester University Press.

Lohor, S. A. (2012). Pre-Colonial warefare among the Chadic-speakers of Jos Plateau and their neighbours (1800-1960). In O. Adeshina, O. Ogen, & N. E. Attah (Eds.), Critical Perspectives on Peace, Coflict and Warfare in Africa: Festschrift in honour of Siyan Oyeweso (pp. 177-186). Ile-Efe: Obafemi Awolowo Press.

Lynch, J. (2008). *Debates in peace journalism*. Sydney: Sydney University Press.

Lynch, J., & MacGoldrick, A. (2005). *Peace journalism*. Stroud: Hawthorn Press.

Macaulay, B. M. (2014). Land degradation in northern Nigeria: The impact and implications of human-related and climate factors. *African Journal of Environmental Science and Technology*, *8*(5), 267–273. doi:10.5897/AJEST2013.1584

Mangvwat, M. Y. (2013). *A history of class formation in the Plateau province of Nigeria, 1902-1960: The genesis of a ruling class*. Durham, NC: Carolina Academic Press.

Marino, E., & Ribot, J. (2012). Adding insult to injury: Climate change and the inequities of climate intervention. *Global Environmental Change*, *22*(2), 323–328. doi:10.1016/j.gloenvcha.2012.03.001

Mazrui, A. A. (2000). Islam between ethnicity and economics: The dialectics of Africa's experience. In T. Salter & K. King (Eds.), Africa, Islam and Development: Islam and Development in Africa-African Islam, African Development (pp. 15–54). Edinburgh, UK: Centre of African Studies, University of Edinburgh.

Mitchell, J. P. (2007). *Media violence and Christian ethics*. Cambridge, UK: Cambridge University Press.

Musa, A. O., & Ferguson, N. (2013). Enemy framing and the politics of reporting religious conflicts in the Nigerian press. *Media. War & Conflict*, *6*(1), 7–20. doi:10.1177/1750635212469909

Nnoli, O. (1978). *Ethnic politics in Nigeria*. Enugu: Fourth Dimension Publishing Company.

Nnoli, O. (2008). *Ethnic politics in Nigeria* (2nd ed.). Enugu: SNAAP.

Ochonu, M. (2008). Colonialism within colonialism: The Hausa-Caliphate imaginary and the British colonial administration of the Nigerian Middle Belt. *African Studies Quarterly*, *10*(2-3), 95–127.

Okidu, O. (2011). One state, many nations: Media portrayal of multiple identities in Nigeria. In L. Oso & U. A. Pate (Eds.), Mass Media and Society in Nigeria (pp. 49-62) Lagos: Malthouse Press Limited.

Onwubiko, K. (1985). *History of West Africa: School certificate. 1800-present day*. Onitsha, Anambra: Africana-FEP.

Oso, L. (2012). Mass media and democracy in Nigeria: The prospect of a Pan-Nigeria public sphere. In M. Mboho & H. Batta (Eds.), The Companion to Communication and Development Issues: Essays in Honour of Prof. Des Wilson (pp. 271-296). Uyo: BSM Resources Limited.

Plateau Indigenous Development Association Network. (2010). *The history, ownership, establishment of Jos and the misconceptions about the recurrent Jos conflicts*. Jos: DAN-SiL.

Plateau State Government. (2004). *Plateau State of Nigeria: Government white paper on the report of the commission of inquiry into the riots of 12th April, 1994 in Jos metropolis September, 2004*. Jos: Government Printing Press.

Plateau State Government. (2009). *Government White Paper Report of the Judicial Commission of Inquiry into the civil disturbances in Jos and its environ, September 2001*. Jos: Ministry of Information and Communication.

Plotnicov, L. (1972). Who owns Jos? Ethnic ideology in Nigerian urban politics. *Urban Anthropology, 1*(1), 1-13.

Rasaq, A. M. (2012). Conflict sensitive journalism and newspaper coverage of Jos conflict. *Entrepreneurial Journal of Management Sciences, 1*(1), 174–188.

Salawu, A. (2009). Media narrative construction of ethno-religious conflicts in Nigeria. *Estudos em Communicacao, 1*(5), 75–93.

Salawu, B., & Hassan, A. O. (2011). Ethnic politics and its implication. *Journal of Public Administration and Policy Research, 3*(2), 28–33.

Samuel, O. O. (2012). *A historical account of ethnic conflicts in Nigeria*. GRIN Verlag GmbH.

Sen, A. (1999). *Development as freedom*. Oxford, UK: Oxford University Press.

Shaw, I. S. (2012). *Human Rights Journalism: Advances in reporting distant humanitarian interventions*. London: Palgrave. doi:10.1057/9780230358874

Shinar, D. (2003). The peace process in cultural conflict: The role of the media. *Conflict & Communication*, *2*(1), 1–10.

Shukla, A. K., & Futrell, J. H. (1993). Collisional activation and dissociation of polyatomic ions. *Mass Spectrometry Reviews*, *12*(4), 211–255. doi:10.1002/mas.1280120402

Steiner, H. J., & Alston, P. (2000). *International human rights in context: Law, politics, morals: Texts and materials* (2nd ed.). London: Oxford University.

Sule, N. R. (2015). *Ethno-religious conflicts, mass media and national development: The Northern Nigeria experience* (PhD Thesis). University of Jos, Jos.

Udoaka, N. (2015). Reporting ethnic minority issues in Africa: A study of Nigerian newspapers. In W. Mano (Ed.), *Racism, Ethnicity and the Media in Africa: Mediating Conflict in the Twenty-First Century* (pp. 117–136). London: I. B. Tauris.

Yusha'u, M. J. (2015). Regionalism and ethnicity in the Nigerian press: An analysis of the coverage of Boko Haram and the Niger Delta conflict in the Guardian and Daily Trust. In M. Wiston (Ed.), *Racism, Ethnicity and the Media in Africa: Mediating Conflict in the Twenty-First Century* (pp. 137–156). London: I.B. Tauris & Co Ltd.

Chapter 2
Clapping From the Arena:
An Enlighten Despotism Using Theatre for Global Development as a Mitigation Against Climate Change on Water Harvest in the Southern Zone of Plateau State, Nigeria

Esther Akumbo Nyam
Plateau State University, Nigeria

ABSTRACT

Climate change is a serious global issue and concern that is attributed to change. A change of climate that is directly or indirectly related to human activity, that which alters the composition of the global atmosphere and which in addition to natural climate variability observed over comparable time periods. There is therefore no doubt that the earth is warming, and the climate changing. Despotism and the rule of despots as agents of democracy has created a rift in the issue of climate change on its citizenry in the southern zone of Plateau State in the area of health, water shortages, cutting meals due to the economic recession in Nigeria. Research has shown that climate change can create a conflict, and it does have a direct effect on scarce resources required to sustain life. Water is at the heart of human existence. Global warming has a major impact on global water cycle, hence on rainfall, soil moisture, rivers, and sea levels. If climate change is not tackled urgently, the calamity will be enormous.

DOI: 10.4018/978-1-5225-5987-0.ch002

INTRODUCTION

Climate change over decades has been a threat globally to environment, health, food, natural ecosystem, tourism, water supply in particular. UNDP (2007) reports shows that over 262 million people were affected by climate disasters annually. From 2000 to 2014, over 98 percent of the people affected are from the developing countries. No doubt the Red Cross since 2006 to date was able to identify that 25 million refugees owed their displacements annually to climate change (Christian Aid, 2006). The issue of water shortage is the focus of this study which needs urgent attention. Water is needed to produce services. Water shortage can be devastating since it leads to drying wells, rivers and the destruction of the ecosystem with the loss of biodiversity and livelihoods, this is synonymous with the southern zone of Plateau. According to Sachs (2008):

evapotranspiration will increase with temperature, so that in hot environments, rising temperatures will make rainwater less available for human use and crop production before it exporates...for some decades, these communities will be threatened by flooding caused by rapid glacier meeting, but after that, the risk will switch abruptly to water scarcity when the glaciers disappear altogether. (P. 89)

The tenet here is the velocity of impact of climate change on water supply and how it poses a massive threat to human life and development and will undermine the international efforts to combat poverty. At this point, rhetorical questions on how the international and national bodies tackle these issues is necessary. It follows with efforts of organisations to tackle the causes, consequences, strategies and implementation of water harvest on the citizens of the southern zone, and enlightenment campaigns for the people of the Southern zone to rise to the challenge through theatre for Development as a tool for sustainable change and development.

WHAT IS CLIMATE CHANGE

The UN (1992) defines climate change as change of the climate that is attributed to human activity that alters the composition of the atmosphere and which is in addition to natural climate variability observed over comparable time periods. Pittock (2009) opines that climate variability is a variability in the average weather behaviour at a particular location from one year to another or one decade to another. This is to say that those changes in the behaviour of the weather over a longer time scales, such as one century to another, is referred to as "climate change".

Research has confirmed that these climate changes are man-made, or anthropogenic. Seitz (2008) concurs that a change in the global climate is now taking place, because of hazardous burning of human large amounts of natural gas, oil, fossil fuels and coal. Carbon is released into the atmosphere as gas, carbon dioxide. Accumulated gases in the air create the green house gas effect.

Global warming is caused by the proliferation of greenhouse gases in the air as the intergovernmental panels on climate change IPCC (2008) are of the view that most of the observed increase in Anthropologic green house gas concentrations is responsible for climate change.

The Paris Agreement (2016) is of importance and beneficial to the continent. In 2016, one of the biggest headlines on the G20 summit was held in China (2016). The announcement of china and United States of America was significant because China and United States of America are the world's two biggest polluters. China is responsible for 20.09% of global emissions and United States with 17.89%. The world Economic Forum's Global Risks report (2016) named it as one of the biggest risk faced by man. In recognition of these reports, 179 countries and European Union spent two weeks in December 2016 in parts to find solutions to the global temperature increase well below 2c and better still below 1.5c. On 11th December 2017, one planet submit was held in Paris with Nigerian president in attendance to address the challenges of the global warming.

The reduction in temperature can only be possible through a significant reduction in the emission of green house gases known as COP 21, (The 21st conference of the parties to the united nation framework convention on climate change) it was one of the largest gatherings of world leaders. The head of states all made emission cutting pledges known as "intended nationally determined contributions" (Alex, 2014). For instance, United State promises to cut pledge on pollution by 26-28% from 2005 levels. China's target is Co2 emissions by 2030 at the latest, then, lower the carbon intensity of GDP by 60% - 65% below 2005 levels by 2030. They head of states also want to increase the share of non-fosil energy carriers of the total energy supply to about 20%. The European Union (EU) plans to cut emissions by 40% by 2030 on 1990 levels respectively.

These findings in Paris on the impact of climate change shows that we are heading for the worst situation if coherent line of action is not taken seriously.

The impact of climate change in Nigeria on water supply varies from regions. Nigeria, like all other countries of Sub-Saharan Africa, is vulnerable on climate change.

The variations of climate change in Nigeria between 1961 and 2000 showed visible occurrences of drought for a cumulative period of 21 years in the Southern zone of Plateau State, Borno, Sokoto, Yobe, Adamawa, Katsina and Bauchi (Muhammed et al., 2011; Obioha, 2009). Water scarcity, drought, rainfall, excessive rising temperature, erosion, and flood are all resultant scenarios that affect man's existence. This explains the migration of man in response to climate change that poses threats to livelihood especially in North Eastern Nigeria.

Similarly, flooding leads to water pollution, soil erosion and destruction of life and properly. The southern zone of Plateau State has witnessed varied degrees of drought, flood which has led to water shortages: erosions, destruction of life and proper as well as farmlands. According to Aluko et al. (2008) climate change has great impact on soil and water supply which is insufficient to sustain the people. At this point, it is obvious that climate change has serious challenges on sustained water supply worldwide especially in the Northern part of Plateau State Nigeria thus, there is need for adaptation on alternative measures of water supply in the southern zone of Plateau State.

Theatre for Development as enlightenment campaign has the ability for conscentization and enlightenment in resolving the problem of water shortages in the northern zone of Plateau State. TFD is about participation, enlightenment, conscientization, change and transformation. There is a wise saying by swami (1992) that: Ignorance is the mother of all the evil misery we see. Let men have light, let them be educated (p.11)

Enlightenment, according to Kant (1946) is man's emergence from his self-imposed ability to use one's own understanding without another's guidance. I am of the view that this nonage is the ability to use one's idea and smartness in taking charge of one's crucial decision in one's life both collectively and individually. This enlightenment should include element such as freedom of expression, nonage, knowledge, communication, people's views and decision for conscientiaztion, enlightenment and sustained development. When a question is asked is the world enlightened? Is there freedom? Is there equal rights? The answer could be a "yes" or a "No". Yet research has shown that the world is enlightened with the computer technology. Not been enlightened either for oneself or others is to violate on the sacred rights of man. Theatre for Global development is a trend in Nigeria theatre today that puts people on the development crossroad in examining and communicating this view, constructive thoughts concerning their issues related to their communities and finding a solution by themselves.

Despotism

Despotism in Nigerian democracy can be likened to the phrase "sheep in wolves clothing". The issue of corruption is the major issue of crisis in Nigeria. This has shown that government as despots are creating a rift in the society and people are now rising to the challenge. Despotism is a form of government in which a single entity rules with absolute power. That entity could be an individual, the despot as in an autocracy, but groups and societies that limit power to specific groups are despotic. The word has a negative connotation. The needs and demands of the common man today is over emphasized and should be interrogated to open room to the people and government for sustainable development.

Theatre for Global Development

Theatre for Global Development is ascribed as a platform for empowerment, creativity, a new idea that gives hopes, at the same time presents an unprofessional response that spurs innovation for collective actions. According to Boon Plashow (2004) said that this kind of theatre develops new industries, creating jobs products and services for quality of life and the economic wellbeing of the masses.

The tenet here is that this unconventional theatre has become a world's money spinner through a creative economy. Kabaso (2013) believes that the theatre, whose objectives is disseminating messages, or to conscientize communities about their set objectives, on socio and economic political situation is vital. This makes the theatre as a tool to build a sustainable and peaceful world. It makes it a powerful and communicative means amongst groups with various problems to voice their though issues through participation by themselves.

TFD is a study of the art form through improvisational process, most times – centred activities. The function of a scripted play is not acceptable and this is what the concept of Forum Theatre stands for. Forum theatre by Augusto Boal, a contemporary playwright deviates from the Aristotle's idea of play who concentrates entirely on the final product. It is largely an act theory that postulates the characteristics values of Arts. The Aristotelian theory is largely based on the theory of Catharsis which is the base and goal of tragic art. Boals Forum theatre destroys the sense of passivity and practically taking control of the drama process as a key to political action the theatre should demonstrate. TF GD being a relatively new movement in theatre can be paralleled to the modern day notions of objectification, reflection and action demonstrated (Brecht & Freire, 1997; Boal, 2000).

The main objective of Forum Theatre is the assumption of putting people first and in practical participation. TFD encourages reversal roles and the people being the focus. The idea here is that it makes it democratic, where both the amateurs and the communities are able to identify their roles and solve them by themselves. According to Kidd (1980), organized groups all over the world, especially peasants and workers, are discovering theatre as a weapon for resolving conflict and other social issues. Forum Theatre belongs to the central concept of popular education which is based on the belief that there are injustices and inequities existing around the world. Forum Theatre includes its acting styles, analysis and transforming the scenario (Frost & Tarrow, 1990). Central to the concept of Forum Theatre is transformation and conscientization; thus a strategy of non formal education. This informal education happens when people who have a common course come together with one voice through collective efforts. 'These efforts of informal education do not happen randomly as formal learning does' (Anderson, 2001). TFD educates people theatrically as a weapon for transformation, change and conscientization by creating issues creatively and interestingly. It means that Forum Theatre is empowering, participatory, educative and interactive. Boal (2000) sees theatre as a forum of knowledge; It should and can also be a means of transforming society (xxxi). This means that theatre should help in restructuring the future rather than the future waiting for it. The use of Forum Theatre is seen as a problem solving theatre that seeks to attack the root causes of the problems and not the symptoms. It is seen to speak the language, idioms and deals with issues that are of direct relevance to the people's lives (Egwugwu *et al.*. 1990). It employs the people's theatrical and dramatic traditions such as songs, dance and folk media.

Furthermore, the role of Forum Theatre by government and non- governmental organizations has become a household name in resolving violent conflict situations in Nigeria. It started in Samaru in 1975 and extended to other parts of Nigerian universities and colleges of education. They are used to teach Nigerian students, NGOs, Governmental bodies use it to mobilize people to talk about their problems. To justify this new quest, new approaches are persistently sought. The strategies of TFD have however proven, most times, to be effective as a tool for transformation, change and conscientization. Conscentization is a popular education and social concept developed by Brazilian pedagogue and educational theorist Paulo Freire, grounded in post-Marxist critical theory which demystify, deconstruct and interrogates the conventional theatre Forum theatre allows participation as a medium for change. The need to begin to look for solution inward is important. This has explained the basis and the role of Forum Theatre practice as an instrument for transformation and sustained development. This study adapted TFD practice in mobilizing the community in harvesting water system by NGOs and government in the Southern zone. This is discussed in the presentation of data and analysis.

Statement of Problem

Over 2 decades, climate change in the southern zone has exhibited visible reoccurrences of drought. The most significant climate change is the continuous persistent decline in rainfall. This shortages connotes metrological droughts that delays rainfall that causes low river discharge and other bodies. Little have been written and contributed by Government bodies to support and encourage these communities that are faced with the challenges of climate change causing water shortages. Activities of depots have become national issue due to corruption and negligence of government in supporting such communities with water projects. There is no proper awareness on what climate change and its effect is and what alternative water supplies should be done to the citizens. This has created a setback on activities of the communities of the northern zone such as looses of food productions, irrigation for farmers, displacement of populations due to migration, declining water resources. The problem is significant because these communities have no access to portable water services. This study therefore sought to access water services and creating awareness through theatre for global development in reaching out to these communities serving as a tool for sustainable development.

OBJECTIVES

1. To access portable water in estimated population of 46,500 people in eight (8) villages from four LGA of the Southern zone
2. To improve the health status of the community members
3. To encourage participation by the set communities into construction of wells and rain water harvest and creating awareness through community theatre practice.

Theoretical Premise

The study employed the Forum Theatre Technique of Augusto Boal (2000). Its ideological context is traced to contemporary playwrights and directors, Edwin Piscator (1893-1966) and Brecht (1964), among others. These scholars sought to create a proletarian drama as opposed to merely producing plays for a working class audience. Piscator's theatre techniques suggested that the stage should supply images and slogans, statistics that would enable the audience to reach political decisions. The essence of Piscator's epic theatre is to learn how to think rather than feel; moving above the steam rather than losing oneself in it (Piscator, 1967). Piscator's Epic theatre is a new direction given to the new modern dramatist and places the audience in control of decision making. It also give the people a voice in the play process. Piscator is of the view that people can rise to the challenge in whatever situation they find themselves. In the same vein, Piscator's pioneering work, Epic Theatre, is now associated primarily with Bertolt Brecht (1964). The Epic Theatre began with the recognition that theatre was growing and moving beyond a simple personification of motive forces or subjecting characters to metaphysical powers against Aristotle's own definition of Epic Theatre. Brecht saw the social relations which created dramatic actions. In other words, social beings condition social thought (Brecht, 1964). Brecht describes this kind of drama as:

A new awareness of selfhood, and people begin to look critically at the social situation in which they find themselves, often taking initiatives in acting to transform the society that has denied them the opportunity of participation (Brecht, 1964).

The issue here is that man is an object of inquiry and centred in a process which points to the fact that society and its structural disadvantages are responsible for the kind of injustice and social inequalities experienced today. To condition man as fixed and inalterable is tragic. It is these contradictions and the ability to deal with them that Brecht discuss. Brecht argues that the artiste should abandon the down stage and move to the community to change it. Similarly, Augusto Boal (1985) is one of the recent theorists of post-modern theatre who provides skeptical, inquisitive, comic and optimistic theatre involving spectators and performance in search of community integrity (Boal, 2000). The essence of this community approach is that the theatre depends so much on the people's culture, as the motivating factor that facilitates social change. The theoretical premise of this study harmonizes the thoughts of Piscator (1966), Brecht (1964) and Boal's (2000) Forum Theatre techniques as a general guide. For Boal (2000), the technique is centred on empowering the communities to take active role in the theatre and society and provoke change through participation. Here, theatre gives room for decision making and active involvement in handling issues which gives the people the opportunity to realize that they form part of their problems and then find ways to resolve them by themselves.

RESEARCH METHODOLOGY/ANALYSIS

The methods of data collection include the Advocacies, JDPC case study reports, focused group discussions (FGDs) semi structured interviews (SSIs) and questionnaires are administered. It concludes with a Drama playlet as a conscientization, and empowerment strategies in encouraging the people into taking action by participating in well constructions and rain harvest projects for sustainable development. The play is achievable through the data that was collected, analysed by descriptive statistics such as tables, diagrammes, pictures, graphs, charts. Qualitative data was achieved through (FGDs).

Presentation of Data and Analysis

The water harvest project was influenced by the growing challenges of climate change on the populace on water shortages in the Southern zone of Plateau State. In view of this, fifteen research teams from the university of Jos, Plateau State University Bokkos and (JDPC) justice development and peace carista set to locate a TFD project in the southern zone of Plateau State.

Preliminary and Logistic Visit

The success of any TFD oriented project has to begin with the preliminary stage. At this stage, the group is able to meet with the various communities to seek permission and discuss issues related to the various communities.

The first preliminary stage was under taken by fifteen staff members and students of PLASU on the 8th and 9th December, 2015 to three LGAs of Quaanpan, Shendam and Mikang respectively. The group was able to see and discuss with the local government chairmen. The chairman highlighted the problems of their communities with burning issue of water shortages and the need to encourage advocacies such as TFD in creating awareness to the various communities. The team also paid a visit to the village head of Ba'ab, in Quaanpan, long Quoemai in shendam and Mikang respectively and they all gave their consent to execute the project.

The second preliminary visit was with a team of fifteen students and members; Prof Illah Sani John, Dr. Nyam, Akumbo Esther. Leading few community members, field officers, (JDPC) justice development and peace caristas members through participatory methodology were used in gathering information.

The delegation on the 15th December, 2015 arrive Shendam town and paid a courtesy visit to the chairman who gave his consent to conduct the research. From the LGA secretariat, four groups were shared for immediate action.

Group A to Quaanpan, Group B Shendam, Mikang and Langtang North was later included as advised by Justice Development and Peace Caritas (JDPC) members. The method used was semi structured interviews, FGDs to achieve data gathering. Flooding methodology was adopted by students and staff in collecting data.

Furthermore, to achieve this, the research phased thus:

1. Phase through to
2. Scenario
3. Analysis identified the problems of the various communities ranging from, health and hygiene promotion, constructions of water supplies such as rain water, wells construction, advocacies and awareness, Hard and bad soil.

Lack of government participation in supplying the populace with water, JDPC as soul implementation of water projects in the southern zone ground and rain water serves the greatest majority in rural area. It has a water supply system of one type or another. The reason is that, among the various sources of supply, ground water and rain water are safe in nature, problems identified are mostly shortages of water hygiene, organizational changes, project implementation.

Rain Water Harvest

In the introduction rainwater is collected when it falls on the earth, it is stored and utilized at a later point. It can be purified to make it into drinking water, used for daily applications and large scale industries.

The Play Analysis

Different scenario are brought together to create a coherent story based on identified problems and suggested solutions through forum theatre techniques. In this technique, we engaged both students and the community based drama group to take part in presentation of analysis through an improvisational play. Their play is titled "Ruwa, shine rai" means "water for life".

Movement one begins with general introduction based on the issue of water shortages in the community. Community members are seen with items going to the stream in search of water. On getting to the stream, they discovered the Fulani herdsmen have gone ahead of them and polluted the water with their cattle drinking water at the centre of the stream. This conflict created a fight that led to loss of life and property.

Movement two: comes a health worker from the community enlightening the community on the dangers of bad water, few women came in bad health and lament on how they got infected through the same source of water they got from the stream. The nurse however, explains the importance of clean water supply and clean environment, the community members went home happy.

Movement three: this movement takes place at the chief's palace. The chief takes his seat as the community members complain on their problems ranging from lack of rainfall, bad water, bad road and so forth. The community also complained bitterly to the chief how the fulani herdsmen had polluted their only source of water supply in the village. At this point, the chief tries to reconcile the two communities on the dangers of conflict. During the question and answer session, people were educated on the importance of quality water supply, alternative measures of water supply, contributing and supporting non-governmental agencies in creating water harvest tanks, well construction and so forth to improve the quality supply of water and its availability for sustainable development in the southern zone. This awareness and conscientization was achieved during the questions and answer sessions. At the end of the play, it became not just a play by the researchers, but for the communities as well. The play allowed exchange of ideas and information between the people and problems familiar to them solving these issues under creative, relaxed and democratic process.

Changes in Organisation

In the past two decades, Plateau State has witnessed changes in the socio - economic and political lives of the people.

The political life and participation of the people were further strengthened by the organization of peaceful Local Government Election across Plateau State. The Local Government system is the closest government agency or organ to the people. That is, through the Local Government, the people of the state can make their voice heard and it is charged with the responsibility to implement impact oriented project at the grass root. In the Nigerian water policy, the Local Government system is charged with greater responsibility of implementing the hard ware component of water and sanitation programmes.

With the incidence of crises and resultant displacement of people from part of Nasarawa state, Justice, Development and Peace/ Caritas Shendam was involved in accessing internally displaced persons (IDPs) in communities of Namu, Pandam, Kurgwi, Lankaku, Bwall, Kwalla, Doemak and Kwa. It is a collaborative work between Quaan Pan and Shendam Local Government Council and NGOS working within the locality. The action has further brought about closer working relationship with the local Government Area of Quaan Pan, Shendam and Mikang and Justice -Development and Peace of the Catholic Diocese of Shendam.

The outbreak of Ebola virus in Nigeria through a Liberian, Patrick Sawyer in July led to nine dead and over twenty affected cases in Lagos, Enugu and Port Harcourt in Nigeria. The outbreak has led to increased awareness and vigilance by Government, Non- governmental organisations and institutions across the country .The personal hygiene level of Nigerians has greatly increased due to increase awareness and jingles on radio and television.

Plateau state through the state Water Board and Plateau Rural Water and Sanitation Agency got a European Union/ UNJCEF grant to improve rural/ urban water supply in two local Government of the state. This will in a long way increase the collaborative network for learning and information sharing to improve project management capacity and service delivery.

Analysis of Set Objectives

The water project was set out with the goal to contribute towards ensuring access to potable water for the local community members and improving the hygiene situation in four Local Government Areas of southern senatorial zone of Plateau state. These will be achieved through self organisation and the participation of the target groups in the construction of water facilities of catchment and improved hand dug wells," and the sustainable use of water which is scarce resources. Two specific objectives were developed to achieve the above project goal.

- **Objective 1:** Access to potable water for a total of 46,500 people in 8 villages of four Local Government Areas (Langtang South 15,000 inhabitants, Quaan Pan: 12000 inhabitants, Pankshin: 11,000 and Shendam: 8,500 inhabitants).
    - **Indicators**
        - 86 wells and 12 rainwater catchments have been constructed.
        - The water quality has improved perceptibly.
- **Objective 2:** The health status of the local community members improved.
    - **Indicator:** By the end of 2014, the number of water related diseases in the target communities has fallen by 10%.

To achieve the project objectives of access to potable water, the project has constructed seventy seven (77) improved hand-dug well and twelve (12) rain water harvesting tanks in rural communities of four Local Government Areas of southern senatorial zone of plateau state and Shendam Diocese. This facility has greatly changed the water quality of communities that benefitted from the projects. The most noticeable are communities of Jigawa, Angwa Alheri, Sabo Gida, Lankaku and kwaikong. Over 50, 000 inhabitants across the communities of southern senatorial zone benefitted from the project activities of well construction and catchment..

In order to improve hygiene and health status of befitting communities, the programme organised and promoted twenty four (24) health and hygiene promotion workshops, it was aimed at promoting change in behaviour towards open defecation; personal hygiene and environmental sanitation for improved health. It facilitates the issue of personal hygiene, open defecation and environmental sanitation is very challenging in the rural communities of the southern senatorial zone. From observation and interactions, the practice of open defecation is more than 60% in most communities across the zone or communities. Also there is the challenge of getting detail statistics of level incidence of water borne or related diseases from community health centres.

ADVOCACY/ AWARENESS AND FOLLOW UP

To enable the successful completion of the remaining projects during the period of six months extension, the project team engaged traditional, religious and political leaders of target communities in a constructive discussions and awareness on the need to assist some vulnerable communities within their locality in payment of local contribution to meet the project implementation requirement. This led to success in Kwa, Bui Jipa Marke. Doka Kaswua, Doka Tofa, Angwa Alheri, Doemak, Namu, Turaki and Tal balong communities.

Figure 1. Activity planned and achievement

S/no	Activities	Planned	Implemented
1.	Advocacy/ awareness Follow	25	24
2.	Health and hygiene promotion	24	24
3.	Well construction	86	77
4.	Rain water harvesting	12	12

Figure 2. Benefiting communities for well construction

SN	District/Community	Villages settlement	No. of Wells	Estimated population
1	Dyis	- Jiblik Permangoes	1	460
		- Jiblik Tinkur	1	550
		- Jiblik Amnat	3	1300
		- Jiblik Aden	1	490
		- Jibilik Jimat	2	1000
		- Jiblik Pustin	1	580
		- Jiblik Labio	1	300
2	Zomo	-Zomo	2	1400
3	Doka Tofa	-Doka Tofa	1	1300
4	Tal Baiong	-Tal Baiong	1	950
5	Doka Kasuwa	- Doka Kasuwa	3	1600
		- Lagurut	1	600
		- Leet	5	2250
		- Nani	3	1500
		- Kwanong	1	450
		- Koplar	1	320
6	Bakin Ciyawa	- Bakin Ciyawa	4	3000
		- Sabo Gida	2	980
		- Sabo Gida Daji	1	250
		- Dawo Daswan	1	300
7	Namu	- Narnu	7	4600
		- Dad in kowa	1	400
		- Lankaku	1	1300
		- Pandam	1	2200
		- Gungkarogom	1	500
		- Tanglong	2	500
		- Salama Dunkun	3	600
8	Kwa	- Kwa	1	3000
		- Meer Kwa	1	270
		- Miket	2	900
		- Kwagar	1	400
		- Varam	1	350
9	Doemak	- Doemak	4	4500
		- Yitlar	1	700
10	Kuka	- Kuka	1	4000
		- lyaka	1	800
		- Saminaka	1	700
		- Jigawa	5	3500
		- Angwa Alheri	1	2200
11	TOTAL		77	**51,580**

Figure 3. Benefiting communities for 46,000 rain water harvest tank

SN	District/Community	Village settlement	No. of catchment	Estimated population
1	Dyis	- Dyis	2	800
2	Jing	- Jing	1	350
3	Kwa	- Bui Jipal - Lardang - Goetoeng	2 1	750 400
4	Kuka	- Jigawa - Merke	1 1	250 350
5	Turaki	- Turaki	1	700
6	Mabudi	- Kwaikong - Mabudi	1 1	400 550
7	Namu	- Namu	1	340
8	TOTAL		12	4890

Figure 4. Implemented activities: Summary of Planed activities and implemented activities for the period April 2014- September 2014

S/no	Activities	Planned	Implemented
1.	Advocacy/awareness Follow	5	7
2.	Health and hygiene promotion	4	4
3.	Well construction	21	17
4.	Rain water harvesting	5	5

Figure 5. Project staff with community theatre group

Figure 6. A visit to village stream of Moedut Village during transect walk

Figure 7. Moedut village stream and inhabitant of Mabudi collecting water at the earth Dam at Mabudi

HEALTH AND HYGIENE PROMOTION WORKSHOPS

In the period of the six month extension covering April to September, 2014, four health and hygiene workshop were facilitated at Marke, Bui Jipal, Angwa Alheri and Kwagar. The workshops drew a total of 97 participants in all the four communities. It was aimed at encouraging behaviour change towards improve personal hygiene environmental care and discourage open defecations in the four villages, through interaction and discussion was observed that, over 70% of household lack access to basic sanitation facility of pit latrines.

WELL CONSTRUCTION

In order to improve and increase sustainable access to potable drinking water in the target communities of southern Plateau state and the Catholic Diocese of Shendam, twenty one wells were planned to be constructed during the six months extension, however seventeen were constructed Namu Tanlong, Kwargar, doemak, Tal baiong, doka Kasuwa, Mere Kwa, Ungwa Alheri, Parmallang, Kalong,

Figure 8. Water project manager facilitating hygiene workshop at Saminaka village

Figure 9. Water point divining at Ungwan Alheri village before well construction

Figure 10. Stages of well construction at Jigawa village of Shendam Diocese

RAIN WATER HARVESTING TANK

To promote improved access to potable drinking water, the project planned to implement five rainwater harvesting tanks of forty six thousand litres (6,000) capacity for challenging communities where, access to safe water is very difficult due to underground and difficult aquifer. Within the extension period, five rain water harvest tanks were constructed in Bui Jipal, Marke, Turaki and Namu communities. The constructions were made possible through support of religious and traditional leaders, and sons and daughters of the communities.

Training and Exchange Visits

The project in the past three years has sent its staff for various trainings and exchange visits. This has helped build and improved the staff capacity to implement the project activities effectively. The project sent staff to Ibadan, Jos, Yola and Jalingo Dioceses for exchange visits, while staff of the programme attended training and capacity

Figure 11. Commencement of catchment construction at Buljipal village

Figure 12. Catchment construction at Marke village at various stages of work

building workshops in the area of peace building, climate change awareness, climate governance and project REDD+, sustainable crop protection, sustainable livestock management, organic farming, strategic planning process, election observation and monitoring, dialogue and mediation for sustainable peace, project management and fund raising.

ADDITIONAL EFFECT AND RISKS

The study as a matter of policy at inception made it clear that gender sensitivity will be main-stream at all level of the project from the management level to the community. In the community, the project emphasized the inclusion of women and youth in the project implementation committee formed at the community level with emphasis on at least 30% women.-Though very challenging, gender mainstreaming is highly advocated by the project.

Figure 13. Climate justice and training climate governance at Hienrich Boll Foundation

Figure 14. Workshop by the theatre groups and JDPC project group

Conflict and peace were experienced during the implementation of the project; the experiences were at various levels. There were internal conflicts as a result of project location, financial contribution for local contribution as experienced in Leet Village, Kwagar, Ampap, K/Ganuwa, Jiblik and Sabo Gida villages. Communal conflict affected project implementations in Kuka, Langtang south and Wase areas. In Jigawa, the project was seen as a source of conflict resolution or peace building.

In the construction of wells, the project work was affected by underground rocks, this lead to non completion of some construction work in villages such as Maraba Demshi, Kwagar, Zomo, Doka kasuwa, Jiblik, Ampap, Tanlong and Varam. The communal crises and sate of insecurity in the state and country have greatly affected access to dynamite used for rock and civil purposes except with special permission, as a programme we are working out an arrangement that will give the program official permission to access dynamites to overcome such challenges.

Impact of climate change is worst affected in farmers crops. Most rural farmers in the southern zone depend on rain for their agricultural productivity. The variability of rain in seasons affects water availability for crop production and domestic use. To address these problems we shall strengthen the project climate change awareness at the community level in other to develop climate change mitigation and adaptation projects.

Civil societies, non- governmental organisation and Governmental agencies have influenced the need for greater networking and collaborations with organisations of like minds. They share common visions and values.

CHALLENGES

Lessons learned within the project. In the course of implementing the project in the past three years, the greatest lessons learnt was that for a project to make meaningful impact, the target beneficiaries should be properly informed and involved in all the stages of the project development and planning. This will reduce misconception and promote acceptance and participation. Some of the challenges in Maikoso community on the nonpayment of their local contribution has affected their implementation of the water project. More campaigns in creating awareness the project is limited and should be encouraged.

Figure 15 Community based drama group in Kwa and plateau state university students presenting their play on the importance of constant water supply

Most rural communities of Nigeria are endowed with natural and human resources that should bring about change in the livelihood and socio-economic conditions of the people but because lack of commitment, passion enlightenment and empowerment the situation has remained unchanged. On the part of government provision of basic amenities such as road, health facilities, water and school will help greatly in the change process.

Because of the low level of understanding and socio-economic condition of most people in the rural areas, development workers must be patient, passionate and committed before any meaningful change can take place.

It was also observed that acceptance, commitment and participation in project activities or change process were dependent on the socio-economic level of the target group in most situations. Communities that participated and benefitted the more have enlightened and empowered individuals in these communities. Also, understanding the seasonal activities of target communities will enhance participation and successes of project activities since most target communities are farmers. It is advisable that project planning and implementation should involve the target community from the beginning to the end.

The study has learnt the strength and capacity of the project as the key to improved and sustained services, implementation at the community level. The ability to network and collaborate with the institutions will enhance and improve the project in a long term.

The New Solutions in Changing World: The Role of Partnership

For sustainable and development of any organization to make any meaningful impact and be world changer, that organization needs to collaborate, network and solicit support from organizations such as Government, non-governmental, traditional leader, political leaders, religious leaders and well meaning individuals. In the last three years of the project, the project had collaborated with CARUDEP, JDPC of the Catholic Archdiocese of Jos. The provincial Justice Development and Peace Caritas, theatre For Development groups and community based theatre groups, Peace and Health, Dadur Bol Development Association had both supported the project in institutional formation and capacity building. There has been support from individuals and politicians who assisted their various communities to meet up with their local contribution. It's a continuous process and on-going.

Planning Monitoring and Evaluation

The major actors involved in steering the project activities are the Diocesan Justice Development Committee who are the project management team, Theatre

For Development team from Plateau State University Bokkos, traditional leaders, religious leaders, other stakeholders such as politicians, government and well meaning individuals target communities are all project implementation committee members. At the inception of the project development phase, the project management team in collaboration with Diocesan Committee involved the communities in needs assessment. The project enabled the research team to identify the basic needs and develop projects that will impact on the life of vulnerable rural dwellers in the Southern Senatorial zone of Plateau State.

At the implementation stage, the project management team engaged the leaders of target communities in advocacy and awareness building and formation of project implementation committee. With this approach, the project team ensured the full participation of the project communities in the project planning, monitoring and evaluation. This will in a long way ensure sustainability and ownership of the project by the target beneficiary communities.

Outlook/Sustainability

In the past three years the project has made appreciable impact on the life of vulnerable rural dwellers of the southern Plateau state in terms of provision of access to sustainable water and sanitation facilities. The level of sustainability in terms of total ownership and sustainability of the project activities by the target community is still at a low level. This is so because without awareness and conscientization of the project, the technical and financial support to most communities and the project activities will not be effective. At the organizational level, the project will still need the support of MISEREOR and other organizations with shared vision to sustain the programme's activities.

Sensibility and Continuation Measures

In order to sustain the activities of the project and increase participation, the project engaged the existing community based theatre Development groups, Parish Justice Development Committee and water sanitation committees. However, observations from most committees were the existence of inaction committees which is a very big challenge .The study encouraged the formation of project water and sanitation committee in the target and benefitting communities. This will enable the project mobilize, implement and, maintain the project and ensure full ownership and participation in project activities. Therefore the formation and strengthening of local community Development associations, Theatre groups, Diocesan Justice Development and Peace Committee will be greatly emphasized to ensure the sustainability of the project activities in the long run.

Local community artisans and labourers are trained during the construction work. By involving and training the community labourers, their capacity and skills are enhanced to be able to construct water facilities for the community and other communities which the project activities does not cover.

The project, to a large extent supports the communities greatly. These are usually through technical support, materials and financial, while communities are enabled to mobilize materials, human and cash to meet up with project implementation conditions. The project was not deliberately designed to hand-over sustainable implementation and continuation of planed measures. Its ongoing which will ensure change of roles where community will be empowered to greatly initiate and implement water and sanitation project themselves.

The project is currently working on developing strategies for generating funding from local and international agencies. However, at the Diocesan level, Justice Development and Peace/Caritas Shendam generate funds through annual Lenten Campaigns. This does not in any way meet the growing developmental needs within the southern zone. At the community level, funds are generated by the local communities through levy, launching and solicitations from well meaning sons and daughters of the community. The funds in most cases are used to meet the cash local contribution required to meet project implementation conditions.

Summary of Analysis

- **Relevance:** From today's perspective, the project is very relevant and important to the target beneficiaries because it enables them access to one of the most fundamental basic needs of life which is water. The technology adopted is the most basic, appropriate and sustainable in terms construction, cost, maintenance and operation, though very challenging when community are not properly mobilized, conscientized and participate.
- **Effectiveness:** The project has considerably achieved its objectives in term of construction and provision of access to safe and potable water, but more need to be done on increasing participation and sustainability of measures when support is withdrawn from the communities.
- **Additional Effects:** The additional effects created by the project in the past three years are many. The project effects will be summarising by a story shared by Mr. Noel Dawam of Lankaku village of Quaan Pan local Government Area. Before the well-constructed in our village, we had a locally constructed well which we manage to construct in my neighbourhood. The well dry up shortly after the rain in November from then onward till the next rainy season when the well will recharge, we trek to Dadin kowa village pond which takes about four hour return trip that will fetch just 20 liters of water for cooking

> and basic chores of plate washing. It is barely enough for personal hygiene and washing. If we want to wash or bath, we all have to trek as a family to the water pond to wash and bath. The collection of water is usually done by my children before they go to school in the morning. In most cases they are late to school. I observe their productivity is low. But all has changed now. With the well water in the community, there has been significant change in my family, my children go to school early, my wife now make local brew which has increased my household hold income and 1 can say personal hygiene has improved. Thanks to the water project.

In the project communities of Dyis and Namu, most of the local labourers and artisans trained by the project in well and catchment constructions avail their skills to private and well meaning individuals that are ready to construct wells or catchment for their household. Their training and empowerment have availed those alternative means of livelihood.

Long Term Impacts

In other to ensure sustainable access to potable drinking water for inhabitants of the target communities of the southern senatorial zone of the state, a proposal to construct 86 improved hand dug wells and 12 rain water harvest tanks is presented to the public during the drama presentation. These measures and option are to ensure sustainable access to water for rural dwellers in the long run. These are the most basic and appropriate water technology that are easy to build and maintain. In most communities where the project is working, access to water from the wells and catchment account for over 50% of basic water use. That is water from the wells account for more than 50% improve water use in the communities.

Sustainability

To ensure sustain ability of measures of the project activities, the project ensured that the benefiting communities are involved from the planning to implementation. The community based theatre groups are formed to ensure continues awareness to monitor evaluate the level of participation of the communities. The people should be involved in the formation of the project implementation, provision of material and resources such as direct labour, sand, gravels, water, accommodation and monetary contributions. These in most cases are done but the total commitment in long term is not very commendable. Communities should be seen owning the process of development, and maintain it without the project interference. This will require patience, commitment planning, training and monitoring over a long period

CONCLUSION

The major financier of the project is the MISEREOR. The MISEREOR, a non governmental body, is a German Aids development agency should be commended for the excellent work that is going on in the northern part of Nigeria. MISEREOR a financial, technical and training supports are of immense value. From the findings, it shows that government is lacking in proper policies, support and implementation of water services in Nigeria. The challenges; the campaign process is a commendable idea that needs to be sustained. Also the initiations of consultancies need to be sustained for greater understanding of work and introduction of new innovations that are impactful and beneficial to the people and the project. With government involvement and support, there will be a difference. Theatre for Global Development has also shown that it has the capacity of bringing people collectively through participation that can be used for sustainable development.

REFERENCES

Alex, G. (2014). *Formative content.* Retrieved from http://www.google.com.ng/sea

Aluko, A. P., Oyeleye, B., Sulaiman, O. N., & Ukpe, I. E. (2008). Climate Change: A threat to Food security and environmental Protection. *Proceedings of the 32nd Annual Conference on Forestry Association of Nigeria.*

Anderson, B. (2005). Violent conflict: Bringing the Real threat into focus, Human Development Report. UNDP.

Anderson, J. (2001). *A popular Theatre/popular Education manual.* Toronto: Cartalyst Centre.

Boal, A. (2000). *Theatre of the Oppressed.* New York: Theatre Communications Groups.

Boon, R., & Plashow, J. (2004). *Theatre and Empowerment: Community Drama on the world stage, case study from around the world.* University of Leeds. doi:10.1017/CBO9780511486166

Brecht, B. (1964). Brecht on Theatre. In J. Willett (Ed. & Trans.), The Development of an Aesthetics. London: Penthuen.

Christian Aid. (2006). T*he Climate of poverty: facts and hope; A Christian Aid Report, May, 2006.* Author.

Egwugwu, E, Steve, O. A., & Lantana, A. (1990). Article. *"NAWE" Journal of the Nigeria Popular Theatre Alliance, 2*(1), 108.

FAO. (2007). Food and Agriculture organisation of the United Nations. In Climate change and Food security: a frame work document. FAO.

Freire, P. (1999). *Pedagogy of the oppressed.* Penguin Books.

Frost, A., & Tarrow, R. (1990). *Improvisation in Drama.* Hampshire, UK: Macmillan.

IPCC. (2008). *Climate Change and Water.* Available from: http://www.icc.ch/pdf/technical-papes/climate-change-water-en-pdf

Kabaso, S. (2013). *Theatre for Development in Zambia.* Kabsy Digital Media.

Kant, I. (n.d.). *What is Enlightenment.* Available at http://www.columbia.edu/acis/ets/ ccread/etscc/kant.h

Kant, I. (1946). *Despotism.* Retrieved from https://archive.org.details.despotis1946

Kidd, R. (1980). People Theatre concretization and struggle in Media Development: Vol. 27. London: Macmillan.

Mohammed, G., Jaliye, M. M., Usman, S., & Baniwa, E. (2011). Adaptive Strategies to climate change variability and change in rural communities. *Proceedings of the 16th Annual National Conference of Agricultural Extension of Nigeria (AESON).*

Nwosu, L. E. (2012). Climate change and food security: Implication for effective agricultural extension services in Nigeria. *Proceedings of the 17th Annual National Conference of Agricultural Extension of Nigeria (AESON).*

Obioha, E. E. (2009). Climate variability, Environmental change and Food Security Nexus in Nigeria. *Journal of Human Ecology (Delhi, India), 26*(2), 107–120. doi:10.1080/09709274.2009.11906172

Piscator, E. (1966). *The Piscator experiment.* Carbondale, IL: Southern Illinois University Press.

Pittock, B. A. ((2009). Climate Change: The Science, Impacts and Solution (2nd ed.). CSIRO Publishing.

Sachs, J. (2008). *Comon Wealth: Economics for a Crowed Planet.* London: Penguin Group.

Seitz, J. L. (2008). *Global Issues: An Introduction* (3rd ed.). Oxford, UK: Blackwell Publishing.

Swami, V. (1922). Complete Works of Wami Vivekananda (Vol. 9). Vedanta Press.

UN. (1992). *United Nations Framework Convention on Climate Change*. Available from: http:/unfcc.int/resource/does/conveng.pdf

UNDP. (2007). Human Development Report 2007/2008: Climate change: Human solidarity in a divided world. Palgrave Macmillan.

Chapter 3
The Impact of Palm Kernel Cake in Resolving Herdsmen-Farmers Violent Conflicts in Ogun State, Nigeria

Abiala Abiala Alatise
University of Ibadan, Nigeria

ABSTRACT

The study investigated the impact of palm kernel cake (PKC) in resolving herdsmen and arable farmers' violent conflicts in Ogun State, Nigeria. There is mounting evidence of violent conflicts, killings, and raping by herdsmen attacks in Ogun State communities. The study was carried out to examine the level experiences of violent conflicts, losses, gains (if any), and the impact of PKC in resolving further violent conflicts among arable farmers and herdsmen. A total of 215 registered poultry farmers were purposively selected which consisted of all the local government areas in Ogun State. A 40-item questionnaire was designed and used to collect information from the respondents. The data collected were analyzed using descriptive statistics such as frequency counts and percentages while inferential statistics such as chi-square were used. Findings revealed that there was significant relationship between the socio-economic characteristics of the respondents and their level experiences of violent conflicts.

DOI: 10.4018/978-1-5225-5987-0.ch003

INTRODUCTION

The Fourth Nigerian Republic's which started in 1999 recorded another history of conflict between farmers-herdsmen violence which killed thousands of people and displaced tens of thousands more. It followed a trend in the increase of farmer-herder conflicts throughout much of the western Sahel, due to an expansion of agriculturist population and cultivated land at the expense of pasturelands; deteriorating environmental conditions, desertification and soil degradation; breakdown in traditional conflict resolution mechanisms of land and water disputes; and proliferation of small arms and crime in rural areas. Insecurity and violence have led many populations to create self-defence forces and ethnic militias, which have engaged in further violence. The majority of farmer-herder clashes have occurred between Muslim Fulani herdsmen and Christian peasants, exacerbating ethno religious hostilities. These conflicts between ever more sophistically armed groups claimed hundreds of lives in 2016. According to the Global Terrorism Index (2010), Fulani militants were the fourth deadliest terrorist group in 2014, using machine guns and attacks on villages to assault and intimidate farmers. After killing around 80 people in total from 2010 to 2013, they killed 1,229 in 2014. Most deaths occurred in the Nigerian Middle Belt, in particular in the states of Benue, Kaduna, Nasarawa, Plateau and Taraba, which recorded 847 deaths. The state of Zamfara, in the northern belt, recorded 229 deaths. In addition to terrorist attacks, Fulani militants were also involved in non-state armed conflicts with groups from Eggon, Jukun and Tiv farming communities. These conflicts resulted in 712 deaths. It reveals that the recent invasion by pastoralists into the oil-rich Niger Delta region of Nigeria has altered the peaceful milieu of the agrarian community, thereby leading to a gradual collapse of Agadama's subsistence base and impeding the growth of sustainable agricultural development in the community. New dimensions to the pastoralists–farmers conflicts in the Niger Delta include the rape of female farmers and shooting anyone found spraying chemicals on grasses/weeds that serve as pastures for cattle. It concludes that the government intervention has been reactionary. However, the role of agricultural extension in resolving conflict between farmers to farmers and farmers-herdsmen can never be thrown over board (Adesiji et al., 2010). In addition, poverty alleviation can be maintained through sustainable root and tuber crops production. (Akele & Chukwu, 2004).

Palm kernel cake (PKC) is a by-product of palm oil production. Palm oil is the world's most widely produced oil, and growing global demand has naturally increased the supply of PKC. While India, Europe and China are the major palm oil importing countries, the bulk of PKC exports go to New Zealand and Europe. A significant amount of PKC is used domestically as cattle feed in Malaysia and Indonesia, where it is fed to feedlot cattle at very high levels. One report states that "it is a common practice in Malaysia to produce complete feeds based on PKC, either as pellets, cubes or total mixed ration." PKC is most commonly produced by economical screw press, less frequently via a more expensive solvent processes. The most useful is palm kernel cake, which is the solid residue left behind after the extraction of oil from the kernels of the palm fruits. It is now well entrenched as a major feed ingredient in beef and dairy feed in the country. The PKC is obtained out from two stages of oil extraction from the palm fruit. Palm kernel cake is a high-fibre, medium-grade protein feed best suited to ruminants. PKC is widely known in West Africa as a viable feed ingredient, but until recently it has been mostly used as a source of energy and fiber in poultry, pig and fish rations. PKC has traditionally been used by small holders as complimentary cattle feed source, its deployment in large herd cattle feed is a recent development. Experts believe that PKC has the potential to make a nation self-sufficient in beef production, and thus solve the major headache associated with dependence on Fulani herdsmen. PKC is gaining prominence as a foreign exchange earner and solution to a major problem in Ogun State since there is inadequacy of grassland resources to sustain domestic cattle production during drought.

Generally, the impact of PKC in resolving Herder-farmer violent conflict can never be over-emphasized. PKC has the potential to solve a major problem associated with beef supply in Ogun State. Most of Nigeria's beef consumption is supplied by semi-nomadic Fulani herdsmen, originally from the northern parts of Nigeria, who can now be found all over Ogun State communities. In recent times, they have been involved in often violent clashes with farmers in towns and villages where they have driven their cattle, destroying crops and water sources. There are regular reports of criminal behaviour among the herdsmen. In April 2016, the Nigerian president ordered a crackdown on Fulani herdsmen who had killed scores of people across the nation. In Ghana, farmers and residents of various communities have often demanded action by government against armed Fulani herdsmen who have regularly destroyed crops, polluted water sources and committed various criminal offenses. And in March 2016, 17 people died when Fulani herdsmen clashed with crop farmers in the north-eastern town of Bouna in Cote d'Ivoire. Governments are seeking to establish ranches as means of solving this problem.

There was an argument that the government, herders and farmers can solve these violent conflicts by deploying palm kernel cake in addition to their limited grassland resources. The viability of PKC, they state, is already proven, and it can be produced in vast quantities for local cattle production, and to increase export revenue. It is an interesting fact that, for many years, West Africans have used PKC and other agricultural by-products in dry season fattening of cattle in small feedlots, where cattle are fattened for 90 to 120 days. This is done to increase weight gain, carcass quality and carcass yield. Although PKC supplies both protein and energy, it is looked upon more as a source of protein. PKC by itself is a medium grade protein feed and with its high fibre content it is often consider as suitable for feeding of ruminants. PKC was ranked a little higher than copra cake but lower than fish meal and groundnut cake especially in its protein value (Devendra, 2007). The nutrient of PKC composed of 16% fiber, palm kernel cake also has a high phosphorus-to-calcium ratio and contains such essential elements such as magnesium, iron and zinc. The typical ration formulated for the feeding of dairy cattle consists of palm kernel cake (50%), molasses (5%), grass/hay (42%), limestone (1.5%), mineral premix (1%) and salt (0.5%) and trace element/vitamin premix. Dairy industry also benefits from the use of PKC when governments determined to stop the cross-territorial herding of cattle, the problem of sourcing feed for the big ranches and feedlots can be solved by the use of PKC. There are substantial amounts of PKC at palm oil processing sites, which would be enough to support feeding at the ranches and feedlots. With plantation farmers accounting most of the growth in palm production, West Africa can count on sufficient amounts of PKC to increase export revenues, and to support domestic cattle production. Indeed, hopes are being raised that PKC could also give West Africa's struggling dairy industry a badly needed thrust. A substantial amount PKC used within Malaysia and Indonesia is used to feed dairy cows. Yet in Ghana, dairy production is based on imported bulk milk products, while Nigeria imports 75 percent. Though West Africa has struggled with the problem of meeting domestic meat requirements, and experts believe that PKC has the potential to resolve herdsmen-farmers violent conflict in Nigeria.

STATEMENT OF THE PROBLEM

Livestock and arable farming are a sub-sector of agriculture which contributes significantly to the economy of rural communities and the nation as a whole. A large percentage of the rural people in Nigeria satisfy their subsistence needs through livestock production involving rearing and marketing of livestock and its products (Okoli, 2004). While livestock includes cattle, goat, sheep, pigs and poultry, reports by Idire (2007) revealed that poultry accounts for about 60% of the livestock production in Nigeria. The arable crop production is also another source of livelihood in which it provides food, income, employment, raw materials etc for the populace. Herdsmen and farmers are very major agents of positive change in agriculture and their contribution can never be withdrawn from the national development. Therefore, the issues of violent conflicts regularly occurring among them posed a major problem and alarming danger on lives and properties in any attacked community in Nigeria. Adetola and Johnson (2017) reported in a print media how women rural farmers beg Federal Government to intervene in cattle herders' attacks (Benue group raises the alarm of Fulani attack). A lot case of herdsmen and farmers' attack claiming thousands of lives and properties had been reported and printed in the media. More so, some other countries like Malaysia and Indonesia have long been feeding their cattle with PKC. Therefore, there is need to investigate the impact of PKC in resolving Herdsmen –farmers violent conflicts in Ogun State, Nigeria.

Objectives

The major objective of this study was to investigate the impacts of palm kernel cake in resolving herdsmen-farmers violent conflicts in Ogun State, Nigeria.

The Specific Objectives were to:

1. Determine the socio-economic characteristics of the arable farmers in the study area.
2. Compare the losses and gains (if any) experienced by arable farmers and herdsmen during and after violent conflicts in the study area.
3. Ascertain the nutritive value of PKC in livestock production.
4. Establish the impact of PKC in resolving arable farmers and herdsmen violent conflicts in the study area.

SIGNIFICANCE OF THE RESEARCH TO THE NATION

This study is an important eye opener to the respondents who are herdsmen and farmers on the need to feed their cattle, pigs, goats, and sheep with PKC. Policy makers and stakeholders will also find the study useful for making policies that will help the production of PKC for herdsmen and farmers available at lowest cost possible. Prospective herdsmen in the local government will find the outcome of the study useful for business decision making before venturing into the cattle business. This study is also important as it will help the Heads of herdsmen in the Local Government Areas or communities to make useful decisions on how to establish peace between herdsmen and farmers within and outside their communities and the extension agents in order to improve their quality relationship. Also, it will help Local government, farmers' associations and veterinary council to organize workshops, conferences and seminars to educate herdsmen about the use and importance of PKC in raising livestock. This study will boost the love, peace and harmony existing in the communities full of herdsmen and farmers. This study may strengthen the workforce of veterinary doctors, extension agents, arable farmers, herdsmen and academic researchers for a peaceful and violent free society.

METHODOLOGY

The study area is Ogun state and is one of the 36 states of the Federal Republic of Nigeria. It was carved out of the defunct Western state on the 3r" day of February, 1976, and it has total land area of 16,409.26sq.km. The estimated population is 3.728 according to Nigerian 2006 National Census released (Federal Republic of Nigeria [FRN], 2009). The climate of Ogun state follows a tropical pattern with the raining season starting about March and ending in November, followed by dry season. The mean annual rainfall varies from 128mm in the southern parts of the state to 105mm in the northern areas. The average monthly temperature ranges from 23 C in July to 32 C in February. The northern part of the state is mainly of derived savannah vegetation, while the central part falls in the rain forest belt. The southern part of the state has mangrove swamp. The geographical landscape of the state comprises extensive fertile soil suitable for agriculture, and savannah land in the north western part of the state, suitable for cattle rearing. There are also vast forest reserves, rivers, lagoons, rocks, mineral deposits and an oceanfront. The rivers in the state provide veritable opportunities for farmers to access the potentials of dry season as well as FADAMA farming. The state capital is Abeokuta, which is about 100km north of Lagos, Nigeria's business capital. It falls between longitude 2° 40E - 6° 40E and latitudes 4° 40N - 9° 15N (Onakomaiyat et al., 2015). Rainfall ranges from 900m in the northern parts, up to 16000mm along the coast. The annual average temperature throughout the year ranges from 21°C - 32°C. The soil type in the study area is ferrasols. The state is made up of twenty local government areas. These are Abeokuta North, Abeokuta South, Ogun Waterside, Ijebu-Ode, Ijebu-North, Ijebu-East, Ijebu-South, Ijebu-North East, Odogolu, lkenne, Sagamu, Obafemi-Owode, Odeda, Ado/Ota, Yewa South, Yewa South, Imeko-Afon, Ipokia, Ewekoro, Ifo and Remo North. The majority of the people of the state belong to the Yoruba ethnic group of south-west Nigeria, and they are mainly of Egba, Yewa, Egun, Awori, Ijebu, Remo, Ikale and Ilaje. The greater proportion of the state lies in the tropical rain forest zone with a sizeable feature of guinea savannah in the far northern area of the state. The main occupation of the people of the state is farming which is largely subsistence in scale. The climate and soil of the state are suitable for the cultivation of a wide range of crops. The major food crops include rice, maize, cassava, yam and banana. The main cash crops include cocoa, kolanut, rubber, palm oil and palm kernels. About 83 percent of the arable crops in the state are grown in the Egba and Egbado areas of the state, while Ijebu Remo areas account for the remaining 17 percent (Ogun State Diary, 2017). However, 82 percent of the tree crops in the state are found in the Ijebu and Ijebu-Remo areas of the state, while Egba and Egbado areas account for the remaining 18 percent. The state is known to have a virile Agricultural Extension Programme which comprises of four agricultural zones identified by OGADEP as Abeokuta, Ilaro, Ijebu and Ikenne. Each zone is further divided into blocks and cells for the provision of extension services in the state. In all, there are twenty (20) blocks and one-hundred-and-twenty-six (126) cells in Ogun and each of these is anchored by a Village Extension Agent (VEA) who oversees the activities of farmers in his coverage area, while a Block Extension Agent (BEA) anchors a block by overseeing activities of farmers in the coverage area.

A multistage sampling procedure was employed for this study. First, the arable farmers were stratified into registered and non-registered in Ogun State. Ogun State was divided into six (6) zones for its operations. The six zones are Yewa, Ijebu, Egba, Ota, Remo and Mowe. In each of the zones, list of registered arable farmers was collected and used as the study population. Forty percent of registered arable farmers were selected randomly and this gave a sample size of Two Hundred and Fifteen (215) registered farmers. Data were collected through the use of a forty item questionnaire; Focus Group Discussions (FGDs) and In-depth-interview (IDI) were also conducted to collect qualitative data for complimentary information. Variables measured for this study included socio-economic characteristics of the respondents, comparism between losses and gains after violent conflicts and impact of PKC in resolving violent conflicts between herdsmen and farmers.

RESULTS AND DISCUSSIONS

Socio-Economic Characteristics of the Respondents in the Study Area

On gender of the respondents, result shows that 68.4% of the total respondents were males while 31.6% of the respondents were females. On age of the respondents, result indicates that none of the respondents was below 30 years of age, 27.9% of the total respondents were between the ages of 30-39 years, 27.0% were between 40-49 years of age and 30.2% were between 50-59 years while 14.9% of the respondents were between the ages of 60 years and above.

On marital status of the respondents, result reveals that 11.6% of the total respondents were single, 69. 8% of the respondents were married, 9.3% of the respondents were separated and 5.1% were widowed while 4.2% of the respondents were divorced. On religion of the respondents, result revealed that 49.3% of the total respondents were Christians and 44.2% of the respondents were Muslims while 6.5% of the respondents were Africa Traditionalists.

On educational qualification of the respondents, result revealed that 20.9% Ofthe total respondents had primary education, 32.6% of the respondents had secondary education and 20.9% of the respondents had tertiary education while 25.6% of the respondents had no formal education at all. This shows that majority of the respondents had education and minority of the respondents had no formal education.

On farm size of the respondents, result shows that 8.4% of the total respondents had large farm size and 45.1% of the respondents had medium size farm while 46.5% of the respondents had small size farm. On farming experience of the respondents, result revealed that 18.6% of the total respondents had between 1-4 years of arable farming experience and 34.9% of the respondents had between 6-9 years of arable farming experience while 46.5% of the respondents had 10 years and above of arable farming experience.

On monthly income of the respondents, result revealed that 49. 8% of the total respondents earned between N10, 000 - N59, 000 in a month, 4.2% of the respondents earned between N60,000 – N99,000 in a month, 16.7% of the respondents earned between N100,000 – N599,000 in a month and 20.5% of the respondents earned between N600,000 - N999,000 in a month while 8.8% of the respondents earned between N1,000,000 & above in a month.

On household size of the respondents, result revealed that 46.5% of the total respondents had a large household size (> 10 persons) and 20.9% of the total respondents had a medium household size (5-9 persons) while 32.6% of the respondents had a small household size (<4 persons). This implies that majority of the respondents had various household sizes they were caring for through the help of arable farming.

On status in associations belonged to by the respondents, result revealed that 12.1% of the total respondents were executive members in their various associations and 78.1% of the respondents were ordinary members in their various associations while 9. 8% of the respondents were not a member of any other associations except that of Farmers' association in their area. It is worthy of note that a very good number of the respondents belonged to one or two association(s) and few of them held executive positions in their various associations.

Results in Table 2 reveals that religion (p=0.011) and education (p=0.001) are significantly related to their level of experiences in violent conflicts with herdsmen at (level of significance p=0.005) and the contingency co-efficient showed the strength of the relationship between the dependent variable (violent conflicts with herdsmen-farmers) and religion, education are 0.14 and 0.21 respectively. Other listed characteristics that do not significantly relate to their level of experiences in violent conflicts with herdsmen and farmers are sex (0.154), marital status (p=0.243) and social status (p=0.147).

Table 1. Frequency distribution of the respondents by their socio-economic characteristics

Socio-Economic Characteristics	Frequencies	Percentages
Gender		
Male	147	68.4
Female	68	31.6
Age (Years)		
30-39	60	27.9
40-49	58	27.0
50-59Ter	65	30.2
60 and above	32	14.9
Marital Status		
Single	25	11.6
Married	150	69.8
Separated	20	9.3
Widowed	11	5.1
Divorced	9	4.2
Religion		
Christianity	106	49.3
Islam	95	44.2
African Traditional	14	6.5
Educational Qualifications		
No formal education	55	25.6
Primary	45	20.9
Secondary	70	32.6
Tertiary	45	20.9
Farm Size		
Large	18	8.4
Medium	97	45.1
Small	100	46.5
Farming Experience (Years)		
1-4	40	18.6
6-9	75	34.9
10 and above	100	46.5

continued on following page

Table 1. Continued

Socio-Economic Characteristics	Frequencies	Percentages
Monthly Income (Naira)		
10,000-59,000	107	49.8
60,000-99,000	9	4.2
100,000-599,000	36	16.7
600,000-999,000	44	20.5
1,000,000 and above	19	8.8
Household Size		
Large	100	46.5
Medium	45	20.9
Small	70	32.6
Status in Association		
Executive Members	26	12.1
Ordinary Members	168	78.1
Not a member	21	9.8

Source: Survey, 2017

Table 2. Chi-square table showing the significant relationship between respondents' socio-economic characteristics and their level of experiences in violent conflicts with herdsmen

Socio-Economic Characteristics	X^2 value	P-value	Contingency Coefficient	Decision
Sex	2.030	0.154	0.068	Not Significant
Marital status	4.178	0.243	0.097	Not Significant
Religion	8.995	0.011	0.141	Significant
Education	19.645	0.001	0.207	Significant
Social status	0.093	0.147	3. 831	Not Significant

Source: Survey, 2017

Other socio-economic characteristics in their actual state of interval of measurement were tested using PPMC. The results in Table 3 revealed that farm size (p=0.010), farming experience (p=0.021), monthly income (p=0.023) and household size (p=0.010) are significantly related to their level of experiences in violent conflicts with herdsmen at (p=0.005). While, age (p=0.681) is not significantly related to their level of experiences in violent conflicts with herdsmen.

Table 3. Pearson Product Moment Correlation (PPMC) analysis between selected socio-economic characteristics of respondents and their level of experiences in violent conflicts with herdsmen

Socio-Economic Characteristics	r-value	P-value	Decision
Age	0.020	0.681	Not Significant
Farm Size	-0.121	0.010	Significant
Farming experience	-0.110	0.021	Significant
Monthly Income	0.073	0.023	Significant
Household size	0.122	0.010	Significant

Source: Survey, 2017

Violent Conflict Experiences of the Respondents

Table 4 presents the results of compared losses and gains of the respondents after violent conflicts. It was discovered that 95.0% of the total respondents experienced loss of farm yield, 97.1% of the respondents had a loss of income/profit, 67. 4% of the respondents reported a loss of family health, 84.3% of the respondents had a loss of quality of peace enjoyed, 68.9% of the respondents had a loss of household resources and 51.6% of the respondents experienced a loss of societal support while 86.9% of the respondents indicated a loss of economic activities.

Categorization of the respondents according to their level of losses and gains experienced after violent conflicts.

The results in Table 5 revealed that 76.3% of the total respondents experienced high losses while 23.7% of the respondents experienced low gains after violent conflicts.

Table 4. Percentage distribution of respondents according to losses and gains when compared after violent conflicts between herdsmen and farmers

Resources	% of Losses	% of Gains
Farm Yield	95.0	5.0
Income/Profit	97.1	2.9
Family Health	67.4	32.6
Quality of Peace	84.3	15.7
Household Resources	68.9	30.1
Societal Support	51.6	48.4
Economic Activities	86.9	13.1

Source: Survey, 2017

Table 5. Percentage distribution of the respondents according to their level of losses and gains experienced after violent conflicts

Categorization	Frequency	%	Mean
Losses (High)	164	76.3	6.0
Gains (Low)	51	23.7	
Total	215	100.0	

NUTRITIVE VALUE OF PALM KERNEL CAKE

Although palm kernel cake supplies both protein and energy, the proximate composition of palm kernel cake suggests that it can be classified as an energy source. The solvent-extracted and expeller pressed palm kernel cakes are of different quality depending upon the extent and efficiency of oil extraction (Sundu et al., 2006). Expeller-pressed palm kernel cake has higher oil content than solvent-extracted meals. O'Mara et al. (1999) reported values of 5 – 12% for expeller pressed and 0.5 – 3% for solvent-extracted palm kernel meal. These two processes thus make the percentage of other nutrients, such as crude protein and minerals, lower in expeller-pressed palm kernel cake. According to O'Mara et al. (1999), the expeller pressed palm kernel cake has lower crude protein content than the solvent-extracted palm kernel cake (Table 6).

With a crude protein content of 14-21% (Table 6), palm kernel cake is precluded as a protein source as in most of the cases, the crude protein level is less than 20%. However, the crude protein content is enough to meet the requirement of most ruminants despite other negative attributes reported by other researchers (Hair-Bejo & Alimon, 1995; Hassan & Yeong, 1999)

The resistance of dietary fibre in palm kernel cake to monogastric digestive enzymes has also been reported by Sundu and Dingle 2003. The fibre content of palm kernel cake (21-23%) has been associated with a decline in nutrient digestibility of palm kernel cake, especially when fed to monogastrics. This notwithstanding, palm kernel cake can still be a major source of energy and protein to farm animals.

Table 6. Proximate analysis of palm kernel cake (g/kg DM) after different oil extraction procedures

Method of extraction	Protein	Ether Extract	Crude fibre	Ash
Expeller	164	78.3	238	49.5
Solvent extraction	197	9.5	268	48.8

(O'Mara et al., 1999)

Table 7. Nutrient composition of palm kernel cake

Fraction	Composition	Reference
Dry matter, %	94	Sundu et al 2005
Crude protein, %	14-21	-do-
Crude fibre, %	21-23	-do-
Lipid, %	8-17	-do-
Ash, %	3-6	-do-
Gross energy, Kcal/kg	4,998	-do-
Metabolisable energy, MJ/kg		
Ruminants	10.5-11.5	Alimon 2004
Poultry	6.5-7.5	-do-
Swine	10.0-10.5	-do-

Analyses have shown that about 49% of the dry matter in palm kernel cake is in the form of nitrogen free extractives (Sundu et al., 2006).

Sundu and Dingle (2003) had earlier reported that during processing palm kernel cake may also undergo Maillard reaction (the reaction of mannose with amino groups leading to the formation of a brown complex) due to heat applied in the process before oil extraction or due to heat evolved during oil extraction, and this will adversely affect the digestibility of the feedstuff.

The fibre content of palm kernel cake may not make it a first choice feedstuff for monogastrics but ruminants with their rumen microbial population can make very good use of the nutrients in them. palm kernel cake has largely been used in Sub-Sahara Africa for cattle feeding, especially dairy animals where it can serve as the main protein source (Bedingar & Degefa, 1990).

Reports by Yeong (1983) and Hutagalung et al. (1982) suggest that the amino acid composition of palm kernel cake is not very good (Table 8). However, very high figures have been recorded for the availability of its amino acids. Nwokolo et al. (1976) had estimates for all the essential amino acids and indicated their availabilities to be in excess of 85% except for valine which is only 68.4% available.

Palm Kernel Cake Utilization in Lactating Cows

Leidiane Rels et al. (2015) concluded in a study carried out on feeding behavior of lactating cows fed palm kernel cake in the diet that the inclusion of PKC in diets for cattle or lactating cows causes no increase in the time spent on feeding and ruminating and does not alter the efficiency of feeding and rumination as well as numbers of

Table 8. Amino acid composition of palm kernel cake, %

	Yeong et al. (1983)	Hutagalung et al. (1982)
Phenylalanine	0.73	0.74
Lysine	0.59	0.61
Histidine	0.29	0.34
Tyrosine	0.38	0.47
Glycine	0.82	0.84
Valine	0.93	0.80
Leucine	1.11	1.14
Isoleucine	0.62	0.61
Threonine	0.55	0.60
Serine	0.69	0.77
Methionine	0.30	0.34
Arginine	2.18	2.40
Cystine	0.20	-
Tryptophan	0.17	0.19

periods and duration of behavioural activities. Maciel et al. (2012) investigated the use of PKC in diet of diary heifers at the inclusion levels of 0.0; 11.9; 22.9 and 34.2% in the diet dry matter and found that the intake of DM and NDF decreased linearly as the PKC was included in the diet. The authors observed that the dry matter intake at the highest level of inclusion of PKC was 54.5% lower compared with the control diet, expressed in kg day. Increased rumination efficiencies of the dry matter and total digestible nutrients may be related to the higher fibre content in the PKC.

Palm Kernel Cake Utilization in Poultry

Due to its high fibre content, the use of palm kernel cake in poultry rations is very limited. Osei and Amo (1987) evaluated palm kernel cake as a broiler feed ingredient. Palm kernel cake partially replaced maize at levels of 0, 5, 7.5, 10, 12.5 and 15%. The addition of palm kernel cake to the diet had no significant effect on feed consumption and body weight up to 8 weeks of age. However, feed conversion efficiency significantly declined as palm kernel cake levels reached 12.5% and above.

Palm Kernel Cake Utilization in Swine

Palm kernel cake has been found to reduce the cost of swine diets. In a study on growing pigs, Okai and Opoku-Mensah (1988) assessed the performance of the pigs on palm kernel cake at 0, 10 and 20% inclusion levels. At the end of the 28-day trial period, all the growth performance criteria were found to be higher for the palm kernel cake diets except for feed intake which was similar for all the three treatments. Back fat thickness was however higher ($P < 0.05$) for the palm kernel cake-based diets. There was a decrease in the cost of a kg feed with increasing levels of dietary palm kernel cake. Rhule (1996) also fed the palm kernel cake at 0, 20, 30 and 40% inclusion levels to grower pigs from 25 to 90 kg live weight without any adverse effect on performance. The average daily weight gains (ADG) by the pigs for the grower stage were 0.46, 0.46, 0.43 and 0.41kg respectively on the 0, 20, 30 and 40% palm kernel cake diets. These values 0%palm kernel cake and 20% palm kernel cake were similar ($P < 0.05$) and significantly higher ($P < 0.05$) than with the 30% palm kernel cake diet, which were in turn similar to those with the 40% palm kernel cake diet. But at the finisher phase, there was not significant difference between the ADGs though that of the 40% palm kernel cake was lower.

In another study, Rhule (1998) evaluated the effect of two diets containing palm kernel cake from two oil mills (A and B), incorporated into the diets at 300 g/kg level. The average daily gains of the pigs were 0.57, 0.46, and 0.49 kg/day for those on control, palm kernel cake A and palm kernel cake B diets respectively, during the grower period. The corresponding values were 0.60, 0.63, and 0.65 kg/day during the finisher period, and 0.60, 0.54, and 0.55 kg/day during the entire grower-finisher period. It was also reported that the palm kernel cake with high level of residual fat led to a higher average daily gain and better feed conversion efficiency but there was increased carcass fatness and reduced leanness in pigs.

Work by Okai et al. (2006) under the AgSSIP Non-ruminant Programme indicated that when a 34.5% palm kernel cake diet was fed to pigs, there was no deleterious effect on carcass characteristics. However, pigs on a control, maize-based diet reached market weight earlier than those on the palm kernel meal diet.

Current Research on Palm Kernel Cake Utilization

It can be inferred from the literature reviewed that palm kernel cake has been accepted as a viable feed ingredient. The only hindrance to its application has been the high level of non-starch polysaccharide which poses a limitation to the rate of utilization.

The current trend in modern animal nutrition is to find means of making the non-starch polysaccharides more available to farm animals, especially monogastrics. One approach that has proved useful in recent years is the addition of enzymes to the feed. The ability of exogenous enzymes to break down high fibre feedststuffs and improve their nutritive value has been reported by Ofuya and Nwajiuba (1990), Chen et al. (1997), Okai et al. (2000) and Iyayi and Davies (2005). Sundu and Dingle (2003) have there are basically three enzymes that are needed to improve the nutritive value of palm kernel meal, namely: mannanase, α galactosidase and celullase to digest the mannan, cellulose and the galactosidic side chains of the palm kernel cake. Laboratory studies have proven this theory (Balasubramaniam, 1976). It is for this reason that exogenous enzymes are now being tested to determine the extent of improvement in the digestion of palm kernel meal.

CONCLUSION AND IMPLICATIONS OF FINDINGS

The research investigated the impact of PKC in resolving herdsmen-farmers violent conflicts in Ogun State, Nigeria. Findings revealed that the religion, educational qualifications, farm size, farming experience, monthly income and household size of the respondents had significant relationship on their level of experiences of violent conflicts between herdsmen and farmers. While, sex, marital status and age of the respondents had no significant relationship on their level of experiences of violent conflicts between herdsmen and farmers. It was revealed that a very good number of the respondents experienced a very huge loss of lives and properties due to violent conflicts. These huge losses experienced by the farmers during and after conflicts contributed to the high level of poverty among the arable farmers. It was discovered that economic activities in communities or villages where violent conflicts between herdsmen and farmers took place may be jeopardized for months and transportation will be hindered, even though, some resident farmers will be forced to evacuate their abodes to another peaceful community or village. However, the major factors of herdsmen and farmers violent conflicts identified on the field with the respondents of this study are over-grazing of farmlands and crops, poor socio-economic status of the arable farmers, trespassing and direct encroachments of the communal and agrarian lands, raping of female farmers, stealing and killing of farmers on the farm when there is very hot argument because the herdsmen moved about with knives, cutlasses, swords, harms and even guns. More so, the impact of PKC either as a supplementary diet or direct feeding for their cattle can never be over-emphasized, there are proven results and testimonies in Malaysia, Asia and some European countries across the globe. The nutritional values of PKC were revealed from past researches and can be used for feeding livestock under ranching which in many ways can reduce or resolving the violent conflicts that claim lives and destroy valuable properties in Nigerian communities among the herdsmen and arable farmers. Palm kernel cake has been accepted as a useful component in animal feeds, especially in Asia and Europe and is also gradually gaining ground in Africa. It is a reasonably good economic feed for both ruminant and monogastric feeding. Since it is an agro-industrial by-product that is being produced locally and within the West African sub-region in sizeable quantities, this feed resource should be fully utilized to reduce feed cost and also curb the problem of environmental pollution that accompanies its disposal. Based on the findings of this study, the following recommendations are hereby proffered:

- There is need for government to make ranching policies and implement them accordingly in all Nigerian communities and culprits should be dealt with irrespective of political connections or power.
- There is need for farmers' associations to organize workshops and seminars to further sensitize the general public, arable farmers' and the community rulers on anger management strategies, escape route and decision making whenever they clash with herdsmen on their farmlands.
- There is need for herdsmen leaders to be sensitized about the impact of PKC, its nutritive values and its incorporation as supplementary diets and whole feed for their livestock as well as ranching methods and coping strategies when grasses or bushes are not available.
- Government should provide subsidies and no-interest credits to the farmers in any affected community in order to help the farmer survive their losses and improve their standard of living at least to reduce their poor socio-economic status.
- Government, security personnel and media should always respond to the farmers whenever they raise alarm about the herdsmen not until after violent conflicts had occurred.
- Herdsmen should be given education and training on the proper ranching and usefulness of PKC as an alternative for feeding their herds instead of violent conflicts.
- Government should encourage more PKC production and PKC industries should be given the necessary assistance they may ever need since they can meet the urgent need of the herds.

REFERENCES

Adesiji, G. B., Akinsorotan, A. O., & Omokore, F. (2010). Farmers Assessment of extension services in Ogun state Nigeria. *Journal of Agricultural & Food Information, 11*(2), 143–156. doi:10.1080/10496501003691661

Adetola, B., & Johnson, B. (2017, October). Women rural farmers beg Federal Government to intervene in cattle herders' attack. Nigerian Tribune. *Tuesday, 17*, 7.

Adisa & Adekunle. (2010). Farmer-Herdsmen Conflicts: A factor analysis of socio-economic conflict variables among arable crop farmers in north central Nigeria. *Journal of Human Sociology., 30*(1), 1–9.

Akele, S. A., & Chukwu, G. O. (2004). *Poverty Alleviation through sustainable root and tuber crops production. In Proceeding of 8th triennial symposium-international society for tropical root crops, Africa Branch* (pp. 1–324). Ibadan, Nigeria: ISTRC AB.

Alimon, A. R. (2004). *The Nutritive Value of Palm Kernel Cake for Animal Feed. Palm Oil Development. No. 40, June 2004*. Malaysian Palm Oil Board.

Asiedu, J. J. (1989). Processing Tropical Crops. ELBS/Macmillan.

Balasubramaniam, K. (1976). Polysaccharides of the kernel of maturing and mature coconuts. *Journal of Food Science, 41*(6), 1370–1373. doi:10.1111/j.1365-2621.1976.tb01174.x

Bedingar, T., & Degefa, G. (1990). *Trends in agro-byproducts and their feeding potential in sub-Saharan Africa.* Working document No. 14, ILCA, Addis-Ababa. Retrieved from http://www.fao.org/wairdocs/ILRI/x5480E/x5480E00.HTM

Chen, J., Lu, T., & Han, Z. (1997). Poultry production in China and the potential for using enzyme preparations. In *Enzymes in Poultry and Swine Nutrition* (pp. 154–155). IDRC Publication. http://www.idrc.ca/en/ev-30911-201-1-DO_TOPIC.html

Federal Republic of Nigeria. (2009). *Farmers led Approaches to Extension. Agricultural Research Extension Network. Abuja.* Overseas Development Institute.

Hair-Bejo, M., & Alimon, A. R. (1995). The protective role of zinc in palm kernel cake toxicity in sheep. *Malaysian Journal of Nutrition, 1*, 75–82. PMID:22692016

Hartley, C. W. S. (1988). The Oil Palm (3rd ed.). Longman.

Hassan, O. A., & Yeong, S. W. (1999). By-products as animal feedstuffs. In S. Gurmit, K. H. Lim, L. Teo, & D. K. Lee (Eds.), *Oil Palm and The Environment: A Malaysian Perspective* (pp. 225–239). Malaysian Oil Palm Growers' Council.

Hutagalung, R. I., Mahyuddin, M., & Jalaludin, S. C. (1982). Feeds for farm animals from the oil palm. In The oil Palm in Agriculture in the Eighties (vol. 11). Incorporated Society of Planters.

Idire, H. E. (2007). *Economics of Demand and supply of vertinary services among commercial poultry farmers in Delta state* (Unpublished B. Agric. Thesis). Department of Agric. Economics, Delta State University.

Iyayi, E. A., & Davies, B. I. (2005). Effect of enzyme supplementation of palm kernel meal and brewer's dried grains on the performance of broilers. *International Journal of Poultry Science*, *4*(2), 76–80. doi:10.3923/ijps.2005.76.80

Nwokolo, E. N., Bragg, D. B., & Saben, H. S. (1976). The availability of amino acids from palm kernel, soybean, cotton seed and rape seed meal for the growing chick. *Poultry Science*, *55*(6), 2300–2304. doi:10.3382/ps.0552300

O'Mara, F. P., Muligan, F. J., Cronin, E. J., Rath, M., & Caffrey, P. J. (1999). The nutritive value of palm kernel meal measured in vivo and using rumen fluid and enzymatic techniques. *Livestock Production Science*, *60*(2-3), 305–316. doi:10.1016/S0301-6226(99)00102-5

Ofuya, C. O., & Nwajiuba, C. J. (1990). Microbial degradation and utilization of cassava peel. *World Journal of Microbiology & Biotechnology*, *6*(2), 144–146. doi:10.1007/BF01200933 PMID:24429985

Ogun State Diary. (2017). *Ogun State land, climate population*. Author.

Okai, D. B., Olympio, O. S., & Anim, P. K. (2000). Responses of grower-finisher pigs to diets containing varying levels of wheat bran with or without Optizyme – an exogenous enzyme complex. *Journal of the University of Science and Technology*, *20*(1-3), 55–61.

Okai, D. B., & Opoku-Mensah, D. (1988). The effects of varying levels of palm kernel meal on growth performance and carcass characteristics of pigs. *Proc. 18-19th Ghana Animal Science Association Symposium*, 54-57.

Okai, D. B., Rhule, S. W. A., Nyannor, E. K. D., & Tandoh, I. (2006). *Feed Packages Based on Local Agro-industrial By-products for Small-scale Pig Farmers in the Ashanti Region*. Paper presented at the Project Leaders Meeting, AgSSIP Non-Ruminant Programme held at Animal Research Institute, Frafraha, Accra.

Okeudo, N. J., Eboh, K. V., & Izugboekwe, N. V. (2005). Growth rate, carcass characteristics and organoleptic quality of broilers fed graded levels of palm kernel cake. *International Journal of Poultry Science*, *4*(5), 330–333. doi:10.3923/ijps.2005.330.333

Okoli, I. C., Anyaegbunam, C. N., Etok, E. B., Uchegbo, M. C., & Udedibie, A. B. I. (2004). Socio- economic characteristics of poultry business entrepreneurs, Imo state, Nigeria. *Journal of Agriculture and Social Research*, *4*(2), 1–11.

Onakomaiyat, D. S. (2015). A Theoretical Extension of the Technology Acceptance Model: Four Longitudinal Fieldss Studies. *Journal of Management Sciences.*, *46*(2), 16–24.

Osei, S. A., & Amo, J. (1987). Palm kernel cake as a broiler feed ingredient. *Poultry Science*, *66*(11), 1870–1873. doi:10.3382/ps.0661870 PMID:3447144

Pimentel, da Silva, Rodrigues Silva, Resende Schio, de Olveira Rodrigues, & de Oliveira. (2015). *Feeding behaviour of lactating cows fed palm kernel cake in the diet*. Retrieved from www. Scielo.br/pdf/asas/37n1

Rhule, S. W. A. (1996). Growth rate and carcass characteristics of pigs fed on diets containing palm kernel cake. *Animal Feed Science and Technology*, *61*(1-4), 167–172. doi:10.1016/0377-8401(95)00934-5

Rhule, S. W. A. (1998). The influence of type of palm kernel cake on the growth rate and carcass characteristics of pigs [Abstract]. *Ghana Journal of Agricultural Science*, *31*(2). doi:10.4314/gjas.v31i2.1928

Sundu, B., & Dingle, J. (2003). Use of enzymes to improve the nutritional value of palm kernel meal and copra meal. *Proceedings of Quensland Poultry Science Symposium, 1*(14), 1-15.

Sundu, B., Kumar, A., & Dingle, J. (2006). Palm kernel meal in broiler diets: Effect on chicken performance and health. *World's Poultry Science Journal, 62*(02), 316–325. doi:10.1079/WPS2005100

Yeong, S. W. (1983). *Amino acids availability of palm kernel cake, palm oil sludge and sludge fermented product (Prolima) in studies with chickens.* MARDI.

Chapter 4
The Business Transformation and Enterprise Architecture Framework Applied to Analyze the Historically Recent Rise and the 1975 Fall of the Lebanese Business Ecosystem

Antoine Trad
IBISTM, France

ABSTRACT

Petro Trad's Lebanese republic was a transitory dominion entity that resulted from and was shaped by the friendly and allied France. The Lebanese-specific cosmopolitan entity was intended to become a Monaco-like structure with a unique, ethnically diverse financial system, with the Lebanese constitution developed to privilege a predominant ultraliberal banking and finance. The idea of a strong financial Franco-Lebanese entity melted down in front of a ruthless ethno-religious antagonistic conflict between Maronite Christians and the Muslim narrow majority that was also fueled by regional and global superpowers. Another very important factor that accelerated the meltdown was the attachment of remote rural and desert regions to the historical Lebanon, or Petro Trad's small Lebanon. The chapter explores the business transformation and enterprise architecture framework to analyze this change.

DOI: 10.4018/978-1-5225-5987-0.ch004

INTRODUCTION

Petro Trad's Lebanese republic was a transitory dominion entity that resulted from and was shaped by the friendly and allied France (World Journals, 2002). The Lebanese-specific cosmopolitan entity was intended to become a Monaco-like structure with a unique, ethnically diverse financial system, with the Lebanese constitution developed to privilege a predominant ultraliberal banking and finance; l system initially designed by Michel Sursock (Tarazi-Fawaz, 2014; Toufic, 2004). The idea of a strong financial Franco–Lebanese entity melted down in front of a ruthless ethno-religious antagonistic conflict between Maronite Christians and the Muslim narrow majority that was also fuelled by regional and global superpowers (Salibi, 1976). Another very important factor that accelerated the meltdown was the attachment of remote rural and desert regions to the historical Lebanon, or Petro Trad's small Lebanon.

The end result was the creation of the Grand Liban (French for *Great Lebanon*), an artificial mixture of 22 minorities who were and still are dominated by the Maronite Christians and Sunni Muslims with a notable influence of the very important Lebanese Jewish community. Nevertheless, against all odds, Petro Trad's Phoenician concept of a financial and business republic survived until 1975 and became the backbone of the outstanding and flourishing Lebanese economy that had its jump start in the '50s. In spite of time and consequent Middle Eastern geopolitical dramas, the Lebanon has preserved its mixed Westernized Phoenician heritage and has even been capable of enclosing other cultures like the Europeans, Arabs, Aramaics, Assyrians, Armenians, and Kurds to make it a unique and a complex culture that in the '70s became the cosmopolitan Lebanese identity. Its exposed cultural and ethnical mixture made the Lebanon a focal point of a networked diaspora and enabled the development of global powerful services backed by a leading financial hub. During the '70s, the Lebanese pound, or lira, became a hard and solid currency, and the Lebanese standard of life became one of the most respected worldwide. Its financial predisposition attracted many institutions and personalities to create and promote their businesses in the Lebanon and its capital, Beirut (Fregonese, 2009). Well known is the case of Beidase's Intra Bank, which became a powerful worldwide bank that enabled the Lebanon to be called the Switzerland of the Middle East and, conversely, made Switzerland the Lebanon of the West. That in turn created tensions, confrontation, and antagonisms between the Lebanon and its Western clone that resulted in the collapse of the Intra Bank and a chain reaction of financial explosions that dismembered the young and naïve Lebanon. The Intra Bank controlled and managed more than 60% of the Lebanese economy. This financial debacle was a blow to Petro Trad's and Michel Sursock's Phoenician concept of a financially based constitution and turned the paradise of Lebanon into a battlefield that destroyed all of its infrastructure and unique cosmopolitan characteristics (Salibi, 1976).

This research uses this historical financial and geopolitical crisis to present a recurrent and frequent scenario of a financial destruction model based on a brutal and sudden societal transformation event that Taleb (2007) would call a black swan. The author uses the case of the Lebanon in an attempt to demonstrate this research's mathematical modelling method and analyzing tool. The author apologizesin advance for any possible offensecreated in discussing these historical and political events. The author basesthis chapter on various sources of information and is aware of the danger that completely opposite views may legitimately exist, whichthe author respects.

BACKGROUND

A societal transformation event leader's role is of crucial importance for the concrete management of a complex societal transformation event and also for its underlying region's or organisation's transformation activities. A societal transformation event leader's or organisation transformer's decisions can be made by using factors within an implemented applied mathematical model. A large set of factors can influence such a mathematical model, including(a) the role of financial law enforcement agencies such as the Swiss; (b) geopolitical factors; (c) political, ethnical and cultural setup; (d) level of education; (e) standard of life; (f) financial competition, which in this chapter is the most important factor; and (g) financial and legal control mechanisms.A globally integrated geopolitical system's approach is optimal to model such events; in this research, it is based on an applied mathematical mode to derive recommendations and explanations (Daellenbach & McNickle, 2005; Trad & Kalpić, 2016).As shown in Figure 1, the decision model interacts with the external world via an implemented research framework using factors which are selected.

MAIN FOCUS OF THE ARTICLE

The Research Process

This research'smain topicis related to Business Transformation Projects (BTP) and the ultimate research question is: "Which business transformation manager characteristics and which type of support should be assured in the implementation phase of a business transformation project?" The targeted business domain is any business environment that encounters: (a) brutal changes; (b) highly complex transformations; and (c) frequent transformation iterations. The author has

Figure 1. The research framework's concept (Trad & Kalpić, 2016)

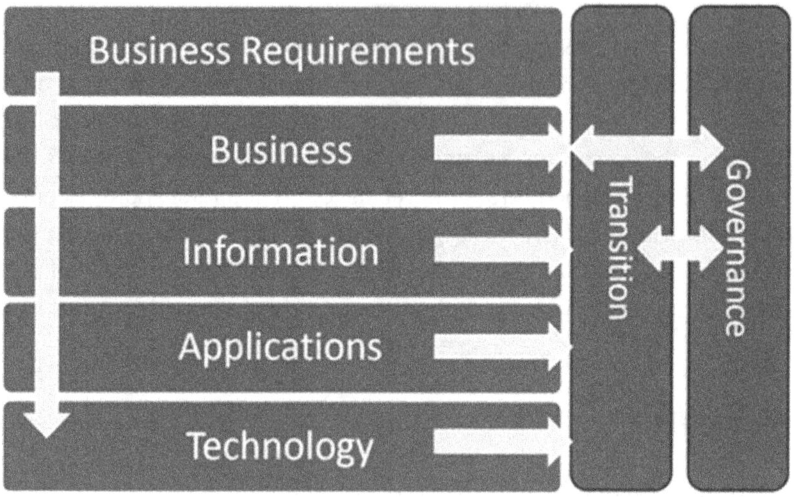

researched the business organisations and their transformation processes with the aim to improve their reason of failure, profitability and chances to survive. Decision making concepts are their main tool in the attempt to achieve this goal. For this goal, they have developed a framework called the *Environment*. The *Environment* or the SmAmCmDmTmPmFmGmF acronym is composed of the following modules:

- **"Sm":** For the selection management of the STEL.
- **"Am":** For the architecture and modelling strategy that can be applied by the STEL.
- **"Cm":** For the control and monitoring strategy that can be applied by the STEL.
- **"Dm":** For the decision-making strategy that can be applied by the STEL.
- **"Tm":** For the training management of the STEL.
- **"Pm":** For the project management strategy that can be applied by the STEL.
- **"Fm":** For the Financial management's support to the STEs.
- **"Gm":** For the Geopolitical mind-mining to the STEs.
- **"F":** For Framework

As shown in Figure 2, in this research the focus is on the Fm, Dm and Gm modules that are bridged by an incorporated Applied Mathematical Model (AMM) that tries to find the reasons of the failure and fatal blow to the very delicate Lebanese financial patchwork and to prove that it is the result of a fierce financial competition; taking into account the global economy (Trad & Kalpić, 2017c; Trad, & Kalpić, 2017d).

This article is a part of a research cluster that has produced many articles, literature reviews, usable items and research artefacts. In this chapter minimal sections of previous works were reused for a better understanding of this complex subject. For this research on the Lebanon, the literature review faced a problem with finding credible academic (re)sources because very little has been done on the subject. Therefore this work can be considered as a pioneer in this field.

For this phase of research, the author tried to deliver their findings and recommendations to clarify the unhappy destiny of the Lebanon and its delicate patchwork. For this purpose this phase's research's sub-question is: "Why did the financial and business Lebanese ecosystem fall?". Where in this case business transformation is focused on banking and tourism (Traboulsi, 2007).

A Delicate Patchwork

The historical, or ancient, Lebanon, situated centrally in Figure 3 in kaki green colour, was the homeland for a Franco–Lebanese entity that melted down into a larger Lebanese republic that at its birth was in a cancerous state, as shown if Figure 3 in grey colour. This state was due to the constant ruthless ethno-religious rivalries between Maronite Christians and the Muslim communities that were amplified and maliciously used by the Lebanon's financial competitors and regional and world powers. The Grand Liban is an artificial mixture of 22 minorities that are dominated by the Maronite Christians, Sunni Muslims, and Shia Muslims with an important

Figure 2. The Financial module interaction (Trad & Kalpić, 2016a)

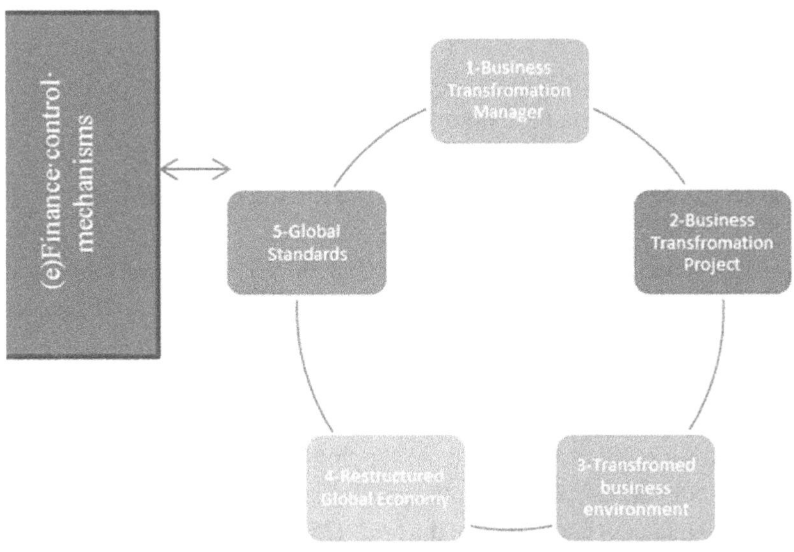

presence, at the country's founding, of the Lebanese Jewish community, all which make the Lebanon a very special and easily inflammable construct because of its long history and ethno-religious heritage (Traboulsi, 2007).

In this chapter, the heritage factor is mentioned due to this country's special religion-based constitution and not for any other reason. The Lebanese ethno-religious maze is not due to any kind of hyper-spiritualty or abundant ethics but rather due to an automatic reflex that draws its origins from the region's socio-political scenery, an unstable and harsh environment. This ethno-religious mirage is a shield that preserves a complex democracy, which is very hard to understand and model but is the nature of the Lebanese entity (Habib, 2009). To simplify this ethnical maze, these various groups can be classified into three major groups: (a) Levantines, or Latins of the Orient, predominantly a Christian population who are a mixture of the remnant of multiple European expeditions and the native Semite Phoenician population; (b) Arabs, the remnant of the multiple Arab Muslim conquests and the converted native Semite Phoenician population; and (c) other groups, including the Lebanese Jewish and Lebanese Armenian communities. What is intriguing in the Lebanese democracy is that each detail causes harsh battles between the 22 authentic ethno-religious minorities. This is a much more complex model than the fairly straightforward Swiss model, in which there are four ethno-lingual groups and Swiss ethnical Germans form the vast majority, with over 75% of the whole population and the country being a predominantly post-Judeo-Christian society (D'Amato, 1995).

In spite of everlasting ethno-religious polemics and tensions, Lebanon's striving communities want to live together in a unique state (United Nations Department of Political Affairs, 1989). This contradicts the reasons for a bloody and destructive 15 years of a senseless civil war that is referred to as the *War of Others* on the Lebanese soil. Antoine Francis's (2002) works prove that an external critical factor for igniting, financing, and supporting a long and fatal civil war in the Lebanon is the real reason for the Lebanese drama.

Critical Success Areas, Factors and Decision Making

Critical Success Area (CSA) is a category of Critical Success Factors (CSF) where in turn a CSF is a set of Key Performance Indicators (KPIs), where a KPI corresponds to a single requirement or feature, as shown in Figure 4, where they are used and managed by the Framework's services layer. For a given problem, an STE analyst must identify the initial set of CSFs to be used for the decision making system. Hence the CSFs are the most important mapping/relation between the geopolitical construct, financial status, organisational predisposition and decision making system; that defines the business ecosystem of the pre 1975 cosmopolitan Lebanon(Peterson,

Figure 3. The historical Lebanon

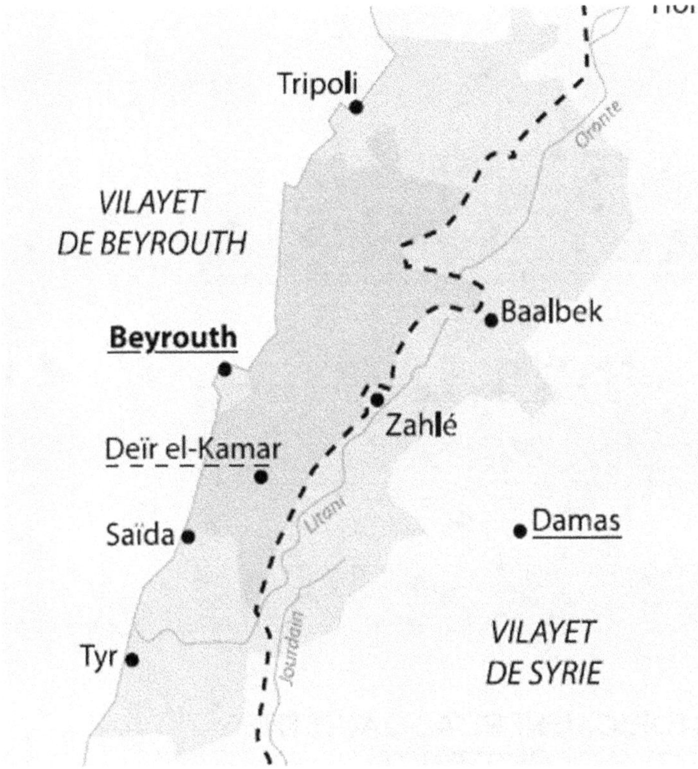

2011; Evans, 2017).The proposed mathematical model or the AMM delivers a set of recommendations and solutions that try to explain the 1975 failure and its real causes(Trad &Kalpić, 2017c; Trad, &Kalpić, 2017d).

The Applied Organizational and Mathematical Model

The proposed AMM is domain agnostic and is not related to any specific methodology and contains a configurable reasoning module that uses CSFs sets and actions(Peterson, 2011); here the enterprise refers to an organisation and the Lebanese entity and its government can be considered as an organization. As shown in Figure 4, the applied AMM supports a micro artefact based architecture, which uses a business-driven approach and uses also the Open Group's Architecture Development Method (ADM)(Trad &Kalpić, 2017b) to manage this article's main blocks: (a)the historical heritage and the geopolitical construct;(b)the financial status;(c)the organisational predisposition; and 4) the AMM and its decision making system.

Figure 4. The Environment's microartefact concept (Trad &Kalpić, 2017b)

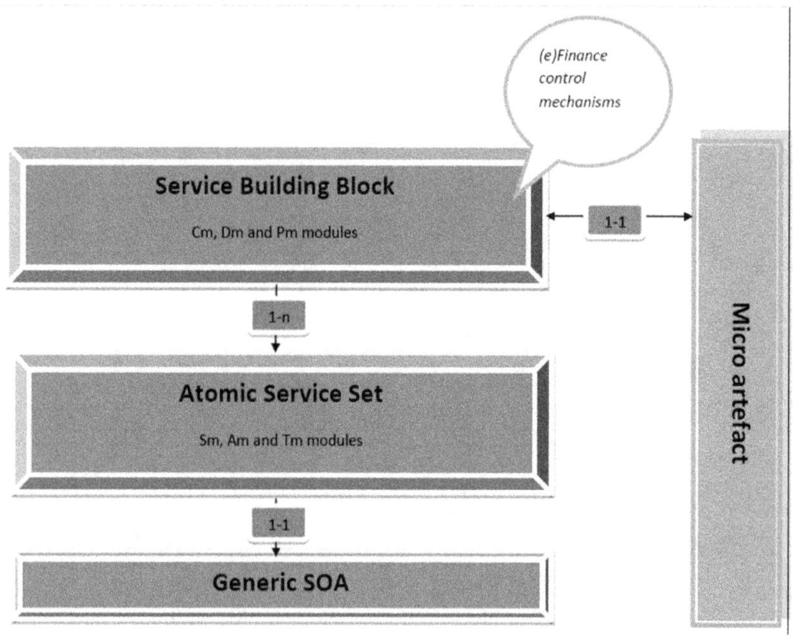

THE HISTORICAL HERITAGE AND THE GEOPOLITICAL CONSTRUCT

This section presents the factors that influenced the bascule of Lebanon in the state of war that destroyed its fine-grained cosmopolitan society that inherited its characteristics from its initial seafaring inhabitants, the Semite Phoenicians. Petro Trad's (World Journals, 2002) Lebanese republic was a French-allied dominion that had a brother-like binding with France, who helped the Lebanon to become a part of the European culture. The Lebanon is a unique cosmopolitan entity that had all the predispositions to become more than simply a competitor of Switzerland (The FactBook CIA, 2018a; 2018b). Lebanon had an ethnical mosaic, tourism, very close ties with the Arab Gulf monarchies, and a loose and liberal banking system that was designed by Michel Sursock. Looking at all the preceding facts, two logical questions arise: (a) Why would the modern and talented Phoenicians cause their own fall? and (b) were they pushed to this fall?

The Phoenician Heritage

Genetic research conducted in the Lebanon and other Mediterranean regions proved that the majority of the Lebanese population carry Semite Phoenician genetic identifiers, what is a common national heritage for the country's various ethno-religious groups (Caquot, 2017). These identifiers are predominant in peoples from the coastal regions. There are three population groups that have impacted the original Phoenician Semite population of the Lebanon: (a) Europeans, mainly French, through various expeditions and military interventions; (b) Arabs, through the progress of Islam that originated from the Arabian Peninsula; and (c) Turkish tribes, through the conquests of the Seljuk Turks,who originated from Central Asia (Zalloua, 2004). The Phoenician nature and affinity for commerce and organized construction that started with Hiram, the Phoenician King of Tyres, remain the main characteristics of the modern Lebanese, one of such is Rafic Hariri, who bewildered many by his ability to rebuild the seriously damaged capitol city of Beirut and other parts of the Lebanon in record time that made Beirut again a regional hub and enabled it to regain its position in the Mediterranean and Arab influence sphere (Favrier, 2017; Zalloua, 2004).

The Arab Influence

The partisans of the honoured Muhammad, the prophet of Islam, drove conquest expeditions, known as *El-Fateh-Al-Islami,* to impose their religious and cultural influence on the Eastern Mediterranean regions. Arab rule and culture penetrated profoundly during the times of the Umayyads and Abbasids, who had a determinant influence on the Eastern Mediterranean regions, including the modern Lebanon, and opened the door to Saudi influence (Hitti, 2002).

The Saudi Influence

The El-Fateh-Al-Islami movement spread the Arab culture in the Lebanon and combined it with the traditional Phoenician curious nature (Zalloua, 2004). This made the Lebanon a leading country in the domains of Arabic literature and culture. That was the binding thread between the original Phoenician population and the Arabian Peninsula incoming tribes that enforced relations with the Kingdom of Saudi

Arabia and other Arabo-Muslim countries. During the Islamic transformational period, the Lebanon became a safe haven for various marginalized minorities. The influx of these diverse minorities created the multiconfessional and multiethnical Lebanese state, where all these groups are represented in government and parliament according to their number and influence (Vloeberghs, 2015). Today, unfortunately, the Lebanese Jewish community is excluded, as it left the Lebanon in the beginning of the 1970s (*Le Parisien*, 2010).

The Lebanese Jewish Community

The Lebanese Jewish community lived in the Lebanon for more than a thousand years and participated in the country's financial and economic fulgurite-like uprise. This community was and still is very much attached to the land of their ancestors. Unfortunately, due to the geopolitical situation in the Lebanon and targeted violence, many of the Lebanese Jewish community members have left the country due to deteriorating security that endangers minorities, including Christians. Very little is done to link the Lebanese Jewish diaspora, or any other minority communities, to the Lebanon; more efforts must be made to warm up and rebuild relations with this very important and historical Lebanese community (Julian, 2014). One Lebanese Jewish citizen residing in New York City in the United States stated, "We are just as proud Lebanese as the rest of the people in Lebanon" (Tabar, 2010, page 1 paragraph 11). The Lebanon has seen the rise of prominent Lebanese Jewish personalities including Jacob Safra, one of the most famous bankers in the world, and Edgar di Piciotto,a Swiss private banking pioneer(Forbes, 2016). The heterogeneous Lebanese society has experienced frequent European expeditions. The effect of this multifold interest has cemented a deeply cultural and human bind between the Lebanon and Europe, and more specifically, France.

The French Influence

The Lebanon and Europe, and more specifically, France, have a geopolitical love affair that goes back more than a thousand years. The Franks, or modern Frenchmen, raided what is today the Lebanon. Surprisingly, chemistry and excellent relations arose between them. This is a singular exception in the rich history of French colonial enterprise. Many Franks settled in the Lebanon and many Lebanese, predominantly Jewish and Christian, settled in France to create a Franco–Lebanese community. This community also includes Lebanese who have never been in France but have fully embraced the French culture, language, and secular republican values; and have even adopted the flag shown in Figure 5, which is the official French flag with incrusted Lebanese cedar.

Figure 5. The Franco–Lebanese flag (Miller, 2014)

The Franco–Lebanese Community

This predominantly Christian and secular community includes many prominent personalities, including Carles Malek, Petro Trad, Charles Helou, Michel Chiha, Camille Chamoun, Charles Debbas, Emile Edde, Amin Maalouf, Carlos Ghosn, and Elie Aboud (Romeo, 2011). The French president and legendary military leader, Charles De Gaulle, who lived in the Lebanon, contributed to a strong vision of the Franco–Lebanese community and relations (Bitar & de Fouchécour, 2015). Following the Franco–Lebanese influence pattern, other communities adjacently developed a similar pattern, including (a) the Saudi–Lebanese community, composed mainly of Lebanese and Saudi Sunni Muslims; (b) the Irano–Lebanese community, composed mainly of Lebanese and Iranian Shia Muslims; (c) the respectable and important Lebanese Jewish community, who participated heavily in the success and the rise of the Lebanese financial system and prosperity of the Lebanon; and (d) a very large cosmopolitan, diverse, and predominantly Anglo-oriented secular layer that gives the Lebanon this special segmented diversity and enriches the Lebanese globally networked lobby. This specific diverse layer needed a long time to mature, took the hardest blow during the civil war, and practically disappeared.

Diversity, Lobbying, and Globalization

The Lebanese main influence communities create a virtual and global multicanton-like federal system that is transnational and promotes a close relationship with many powerful world communities. The evolution of a globalized Lebanon is a financial and transformational concept that must incorporate collaboration with (a) all the region's countries, (b) international institutions, (c) national sectors, (d) the global information network, and (e) local and regional laws. Added to existing international and national laws, there are legal frameworks, conventions, treaties, and directives to be integrated to support this transformed society. The directed 1975 black swan effect damaged the fast and chaotic evolution process of Petro Trad's Lebanon (Taleb, 2012).

1975: The Directed Black Swan

The Prelude

Various important events listed as follows happened in the Lebanon before the year 1975 that where the programmed prelude to the regional and more specifically, the Lebanese, march to the unknown:

- **1913:** Petro Trad, representing diverse Christian streams, demands an end to the Ottoman military occupation of the Lebanon and calls for an independent Franco–Lebanese entity. This actdeeply angered the Ottoman occupier against the Lebanese Christians.For this, Petro Trad and his close colleagues were condemned to death (World Journals, 2002).
- **1915:** The Great Famine strikes the Christians of Lebanon and the Ottoman occupiers seal off their regions. The Christian population was decimated at about 85% where 40% and another 45% fled to France, Brazil, the United States, and other countries (Fersan, 2010). This event changed the demographical structure of the Lebanon, which had been a Christian state in which Christians formed 90% of the population.
- **1926:** The creation of the Grand Liban by France.
- **1943:** Petro Trad becomes the President of Lebanon.
- **1943:** The 1943 National Pact is concluded and independence is declared. This event helped to define the constitution of the free democratic Lebanon and the share of power between the Christian majority (52%) and Muslim minority 90% (45%) on the basis of 6:5 proportions in favour of the declining Christians (Faour, 2007). In this new republic, the dominant Christians were split into three main categories: (a) Lebanese nationalists, mainly Maronites; (b) pro-Syrian nationalists, mainly Greek Orthodox; and (c) other Christians, mainly Greek Catholics.
- **1949:** Pro-Syrian nationalist and Orthodox Christian nationalist, Antoun Saadé, who wanted to embed the Lebanon into the Syrian sphere that resulted in a bloody confrontation with the Christian Lebanese nationalists who in turn wanted an independent and Westernized Lebanon (Solomon, 2017), is

captured and executed. This was a prelude to widening the Lebanese internal conflict and opened the door to the Nasserites' pan-Arab failing intervention.

- **1958:** The 1958 Nasserites Pan-Arab Movement ignores the Lebanese 1943 National Pact and aims to destabilize the free democratic Lebanon. This event ended with the U.S. Marines' intervention that saved Camille Chamoun's newly born Lebanon, caused the Lebanese society to polarize further, and forced Francophone Lebanese Christian nationalists to shift to the Western sphere (Little, 2007). This was the age of rising Arab socialist revolutions that had a bloody culmination in the Algerian civil war.
- **1962:** The Algerian civil war ignites an anti-Western and anti-French sentiment that increases the divide in the Lebanon, which had a similar division scheme in which the two major communities ended being anti- or pro-France. However, the Lebanon continued its financial uprise.
- **1963:** Despite all the previously listed events, the Lebanon becomes a financial power and largely surpasses its Western copy and other competing countries that have similar economies. This uprise was labelled as the *Lebanese economic miracle* that awakened and alarmed Lebanon's competitors (Kubursi, 1999).
- **1966:** Alarmed financial competitors respond and the Lebanese financial giant, Intra Bank, collapses as the Lebanon enters a difficult period. The Intra Bank managed and controlled more than 60% of the Lebanese economy and wealth (Abu-Fadil, 2017). This difficult situation was aggravated by the regional Arab–Israeli conflict.
- **1967:** The Arab–Israeli War comes as a result of the pan-Arab euphoria in which the Lebanon refused to intervene and maintained a weary neutral position. This position was seen as an act of treason by Arab states and ignited tensions between Lebanese Christian nationalists and the dominant Palestinian Liberation Organisation (PLO) that was backed by (a) financial circles in Switzerland, (b) the Lebanese Muslim communities, and (c) pro-Syrian nationalists.
- **1969:** Increasing Arab state pressure and anger ends with the 1969 Cairo agreement that strips the Lebanon of its sovereignty and surrenders the country and its institutions to the PLO. This came as a punishment to the free democratic Lebanon because of its neutral behaviour in the 1967 Arab–Israeli War (Sorby, 2012). At his return to the Lebanon, President Charles Helou assembled main Christian leaders Camille Chamoun, Pierre Gemayel, and Jean Njeim and recommended to arm and expect war(El-Khazen, 2000).

- **1970:** The Palestinians were caught in the Black September War in Jordan that dismembered their main current, the PLO. Due to these events, PLO members escaped to the Lebanon and the 1969 Cairo agreement blocked the Lebanese armed forces from reacting.
- **1970:** Hafez El-Assad becomes the president of Syria.
- **1970:** Suleiman Frangieh becomes the president of Lebanon.
- **1970:** A helicopter crash kills powerful army commander Jean Njeim, a fierce enemy to the PLO expansion in the Lebanon (Lebanese Army, 2017). The Lebanese Army was put in a spectator status and the PLO overran the Lebanon by looting and discriminating the Lebanese population, especially minorities(Dahl, 2006).
- **1971:** Brutality against minorities hits a new peak with the assassination by the Syrian Secret Service of Albert Elia, a Lebanese Jew who was the president of the Beirut Council (*L'Orient-Le Jour*, 2017).
- **1973:** A mini war breaks out between the PLO and Lebanese national army in which the national army is paralyzed by Assad's threats to invade the Lebanon. The Lebanon was forced to respect the 1969 Cairo accord that had opened the door to the planned civil war (Dahl, 2006).
- **1973:** The Arab–Israeli War is proceeded by changes of regimes in the Arab states to socialist republics including Assad's Syria, which has an aggressive military attitude (Aker, 2014).
- **1973:** U.S. secretary of state Henry Kissinger accounts Assad's aggressive ambitions in the rich Lebanon and observes a possible solution to the Middle East confusion. The solution lies in ousting the Christian community and replacing them with the Palestinian refugees (Hijazi, 1975).In this way, the Lebanon would become simply a province in Assad's empire. This scenario was also supposed to bring an end to the Arab–Israeli conflict (El Hashem, 1990). This also meant the end of French interests in the Lebanon and Middle East. Apparently, this was a shortsighted strategy that proved to be faulty as Assad seized the opportunity in 1975 to fill his Swiss bank accounts (The International Consortium of Investigative Journalists, 2015).
- **1973:** Antoine Kamouh, the Intra Bank banker and arms dealer, disappears in Paris. He was the link between Beidas's Intra Bank and the Swiss bankers; his activity was to arm various terrorist groups in the Lebanon (Johnson, 1983).
- **1975:** The Saudi king, Faisal bin Abdulaziz Al Saud, is assassinated. He had a steady hand that protected the Lebanon and wanted to keep it as an independent multicultural entity (Haley & Snider, 2006).
- **1975, April 13:** The final act starts the well-financed 15-year Lebanese civil war(Fregonese, 2009).
- **1975:** In April, the Vietnamese civil war comes to an end.

- **1975, May:** An official World Bank report states, "Lebanon is aiming for European living standards and should be compared with a European country in the lower income bracket" (Toufik, 2004). Underlying this assessment was a remarkable financial performance although it had supported more than 1 million refugees.
- For the Lebanon's main competitor, this was more than a new alarming signal.

The Final Act

Hour zero, on the 13th of April 1975, a carefully planned war submerged all Lebanese regions without any effective will from any side to stop it. The national army became a passive spectator to the orchestrated dismembering of a rich and multicultural modern society. The initial minor incident in April, in which Joseph Abou Assi was assassinated, caused consequences that took 15 years to stop, destroyed the Lebanese economical infrastructure, and exterminated the rising generation. The 1975 civil war was an externally orchestrated deed and contradicts the reasons for a bloody and destructive 15-year-long civil war that is today referred to as the *War of Others* on Lebanese soil. The works of Antoine Francis (2002) proved an external factor of igniting and supporting a long and fatal civil war is the real reason for the Lebanese drama. One solid argument in favour of a planned war is that the Arab and some Western countries (excluding France) ignored the Jordanian army's crushing of the Palestinian movements and prohibited the Lebanese nation from protecting itself from incoming waves of combatants (Francis, 2002). These factors enforce the view that geopolitical elements were moderately important and locally significant, and that finance looms as a major reason for this brutal fall. The behaviour of Assad's Syria towards the Lebanon's drive to the unknown and his Swiss bank accounts may seal the conclusion that the Lebanon civil war had an external transformer.

The Geopolitical Critical Success Factors

Based on the literature review, the most important facts were used as Geopolitical Critical Success Factors which in turn are used in the *Environment's* decision making system. This section's (CSA: Geopolitics) set of filtered CSFs and their weightings are shown in Table 1.

The Modules Chained Link to the financial Reason

As we see in Figures 6 to 15, the financial factors and Assad's Swiss Bank accounts played the most important role in the 1975 drama; whereas the regional powers did not gain any significant geopolitical advantage of the Lebanese saltomortale. This

Table 1. The geopolitics critical success factors

Critical Success Factors	KPIs	Weightings
CSF_Geopolitics_PhoenicianHeritage	Admitted	From 1 to 10. **2 Selected**
CSF_Geopolitics_FrenchInfluence	Priority	From 1 to 10. **5 Selected**
CSF_Geopolitics_SaudiInfluence	Important	From 1 to 10. **3 Selected**
CSF_Geopolitics_1975	Intentional	From 1 to 10. **10 Selected**
CSF_Geopolitics_AssadSwissBankingAccounts	Intentional	From 1 to 10. **10 Selected**

section's deduction is that the 1975 war was intentionally ignited and that there is a need to inspect more the real reasons for this fatal event and to inspect also who really profited from it. As geopolitically nobody really profited from this sad event, the focus will be next on the financial ambitions.

THE FINANCIAL REASON

The applied mathematical model is adapted to model the Lebanese geopolitical and financial problem, and explain the fall of the Lebanese financial ecosystem. The proposed model uses the following types of information sources: (a) banks, (b) business valuation,(c) currency value,(d) global domestic growth, (e) finance institutions, and (f) global financial networks. These sources are needed to construct the model's CSFs.

Banks

The Lebanon's banking and financial environments attracted many regional and international institutions and personalities who were interested in promoting their financial activities. The Lebanon enabled these institutions to get closer to the immense Arab oil-based wealth, the petrodollar dividends of which were pouring into Lebanese bank reservoirs, making the Lebanon a worldwide leading financial power. This dried up the Swiss financial ecosystem and forced it to concentrate on tourism and chocolate production. One of the implicated banks was the Swiss-influenced Intra Bank, founded by Yousef Beidas, which in a short time became a very solvent worldwide bank. Switzerland, which felt it was too far from the Middle East and Gulf region, became weary of its hyperactive and unpredictable Middle Eastern clone and viewed the Lebanon as jeopardizing its financial system. Discovered and then bullied, Beidas escaped from the Lebanon and found a protective new bunker in the Swiss mountains in the region of Luzern. Beidas vanished under mysterious conditions, leaving no information or trace. These facts inflamed tensions and antagonisms between the Lebanon and its Western counterpart. Confrontation and suspicion forced the Lebanese government to liquidate the Intra Bank, which generated rumour attacks that in turn caused a chain reaction that in turn caused financial chaos and social explosions. These events disfigured the young and defenceless Lebanon. Beidas's actions were an intentional blow to the Phoenician concept of an unwritten financial constitution and turned the Levantine paradise into an exhausted battlefield that destroyed Lebanon's infrastructure and human diversity. It also resulted in the collapse of the Lebanon's touristic system (Abu-Fadil, 2017). In fact Switzerland and Lebanon have very similar economies based on private banking (Indexmundi, 2018).

Tourism

Another Lebanese sector that suffered heavily as a result of its civil war is tourism. The Lebanon had a variety of touristic domains, including (a) winter skiing in its snowy mountains, (b) summer tourism, (c) jet-set meetings and nightlife activities, and (d) historical sites and religious tourism. In 1970, the Lebanon numbered 3 million inhabitants and 1 million refugees, and received more than 1 million tourists (Blake & Wagstaff,2016). The 1975 disruption transferred the touristic focus to the neighbouring Cyprus, to which many Lebanese have immigrated and participated in its important touristic evolution, which had an effect on its economic growth, as shown in Figure 6.

Figure 6. Cypriot economy growth (Trading Economics, 2017h)

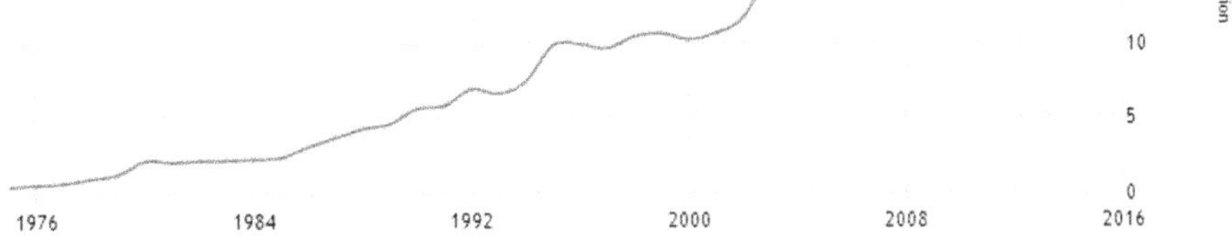

Business and Economic Growth Valuation

Business valuation is a process that includes a set of procedures that can be applied to estimate the value of a business entity. It is a complex undertaking to evaluate a business value; the value to be precisely estimated on the basis of concrete business valuation assumptions. It is not easy to establish the real value of a business environment because business value comprehends many different factors and also depends on financial specialists' views. Business entity stakeholders look at businesses' connection to a country's macrobusiness values (ValuAdder, 2017).The real economic growth rate estimates economic growth, which is related to gross domestic product (GDP), which is adapted to inflation. The real economic growth rate is presented as a percentage that values the rate of evolution for a country's GDP in a given period of time, usually a year (Investopedia, 2017a).

Economic Growth of Regional Powers

This section inspects if regional powers including Syria and Israel took advantage of the Lebanese tragedy. From a financial point of view, the Lebanese tragedy did not influence neighbouring regional powers including Israel, Syria, or Iran, as shown in Figures 7 and 8; where we see their respective GDPs. In the end, these regional powers did not gain significant financial advantage.

Figure 7. Syrian economy growth (Trading Economics, 2017b)

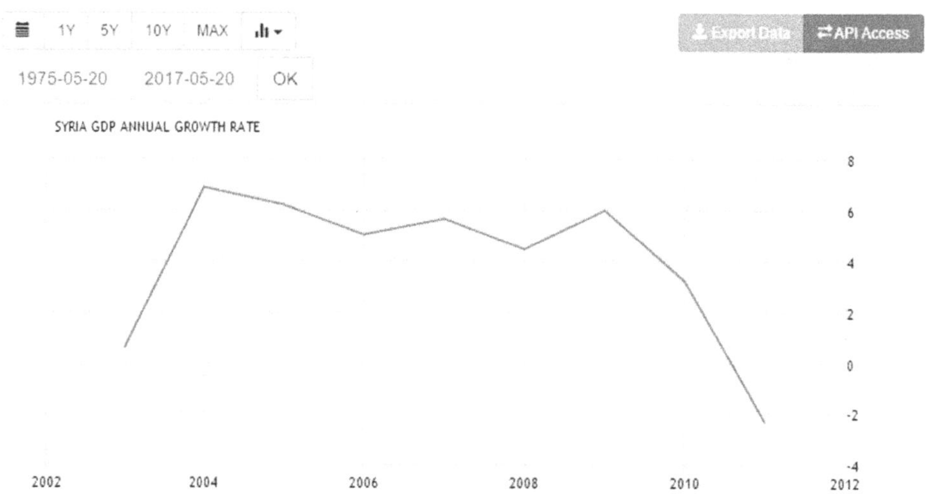

Economic Growth of Western and Global Powers

This section inspects if global powers GDPs, including the United States, the United Kingdom, and France took advantage of the Lebanese civil war. As seen in Figures 9, 10, and 11, financially from the GDP's, these global powers did not acquire any significant advantage.

Economic Growth of Switzerland

This section inspects if Switzerland took advantage of the Lebanese civil war, by looking and comparing GDP diagrams. As shown in Figures 12 and 13, analysing the GDP diagrams' slopes and it seems obvious that the Lebanese declining GDP slope is inversely equivalent to the Swiss financial takeover. Switzerland is the unique country that gained financial advantage of the Lebanese tragedy, with the Lebanese GDP slide weakening its national currency.

Currency Usage Strategy

At the beginning of the war, the Lebanon had its GDP drop to one third of its GDP of the early '70s. Although its currency before the war evolved to become a worldwide solid currency, there were speculations on replacing the legendary petrodollar with the Lebanese petrolira. Lebanon's fast enrichment placed it in the shooting range of the financial hunter of the Swiss and enabled the Lebanon's replacement with its Western clone (Rolland, 2004). The rumour-based collapse of the Intra Bank and

Figure 8. Israel's economy growth (Trading Economics, 2017c)

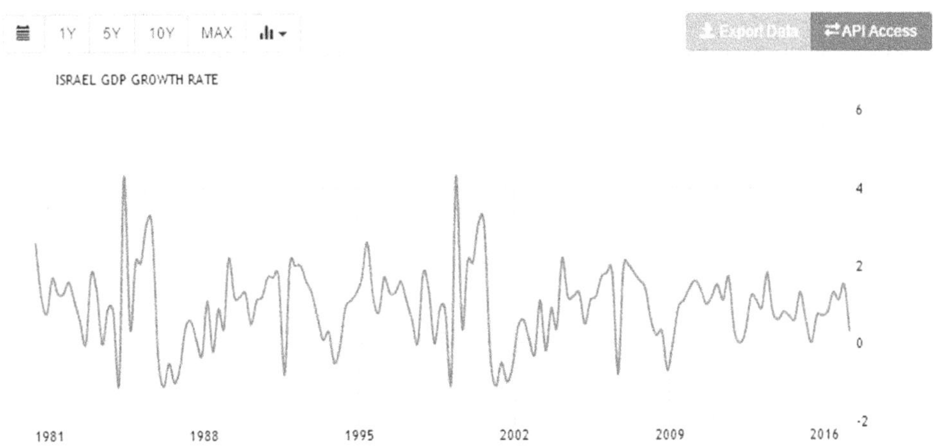

Figure 9. U.S. economy growth (Trading Economics, 2017d)

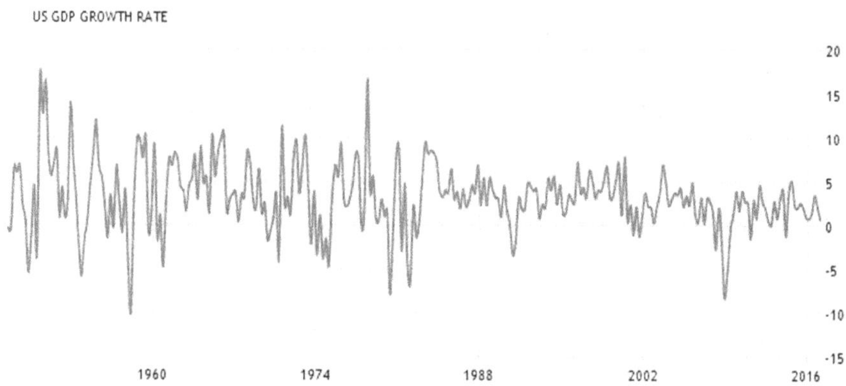

Figure 10. France's economy growth (Trading Economics, 2017e)

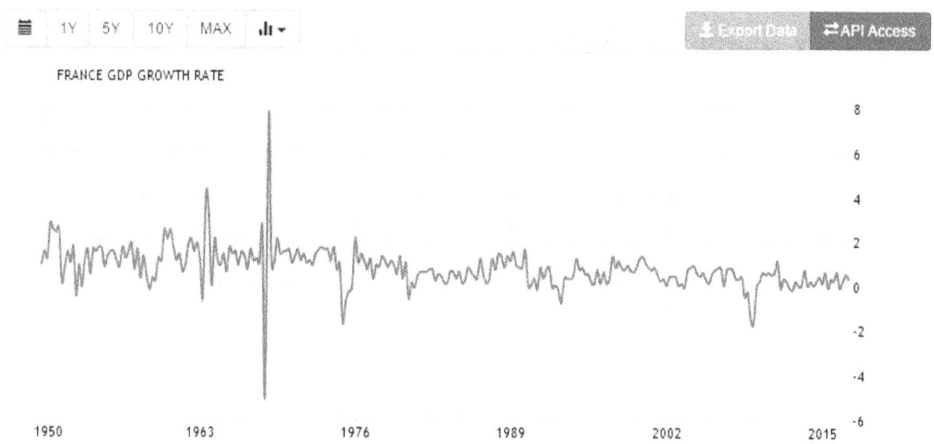

other Lebanese financial giants caused the migration of financial assets from the Lebanon to its Western equivalent (AMInfo, 2014). As shown in Figures 14 and 15, the Lebanese currency was replaced with the Swiss hard currency. More precisely, the Lebanon was replaced by its unique competitor, the Swiss Helvetic Confederation.

The Swiss Competitor

To better understand the reasons of the 1975 Lebanese fall, the author researched the Swiss collective behaviour and found the following:

Figure 11. U.K. economy growth (Trading Economics, 2017f)

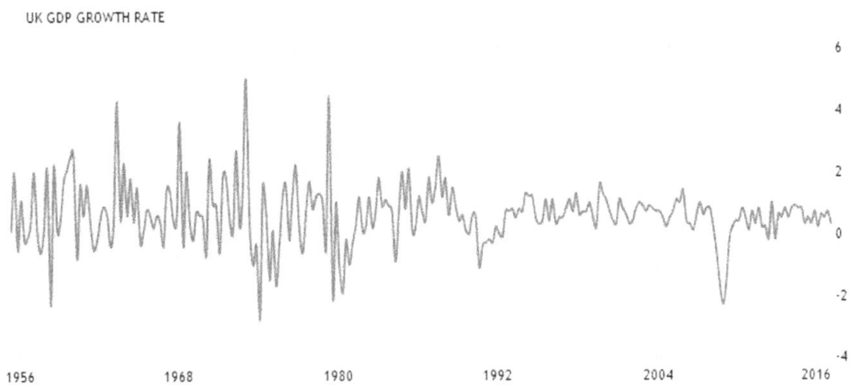

Figure 12. Switzerland's economy growth (Trading Economics, 2017g)

- Switzerland is very similar to the Lebanon and has practically an identical financial model that is based on Gulf countries' wealth management.
- Switzerland's main political and ruling party, the SchweizerVolksPartie, is an extreme far-right racist and anti-Semitic fascist entity that has a 1933-like attitude (Maurisse, 2016; Miller, 2017; Tribune de Genève, 2014).
- Unlike the Lebanon, Switzerland is not a cosmopolitan state (Toufic, 2004). Only the canton of Geneva can pretend to have such a status but the other Swiss cantons can be considered xenophobic. Also, the Lebanon had and still has one of the highest literacy rates in the world (Toufic, 2004).

Figure 13. Lebanon's economy growth (Trading Economics, 2017a)

Figure 14. Switzerland's currency evolution (Trading Economics, 2017i)

- The Lebanese is a laissez-faire economy whereas the Swiss one is a precise and organized one (Toufic, 2004).
- Switzerland and its financial institutions have been the main leaders in worldwide financial scandals, misdeeds, and criminal acts including (a) the Libor manipulation, (b) currency manipulations, (c) credits manipulations, (d) supporting arms dealing transactions, (e) hijacking people's wealth, (f) financial fraud, (g) the subprime crisis, (h) war victim wealth confiscation, (i) tax evasion, (j) drug dealing financial support, (k) war support for future financial competitors, and (l) arms dealing(Clarke &Tigue, 1975; Parker, 2016).

Figure 15. Lebanon's currency evolution (Trading Economics, 2017j)

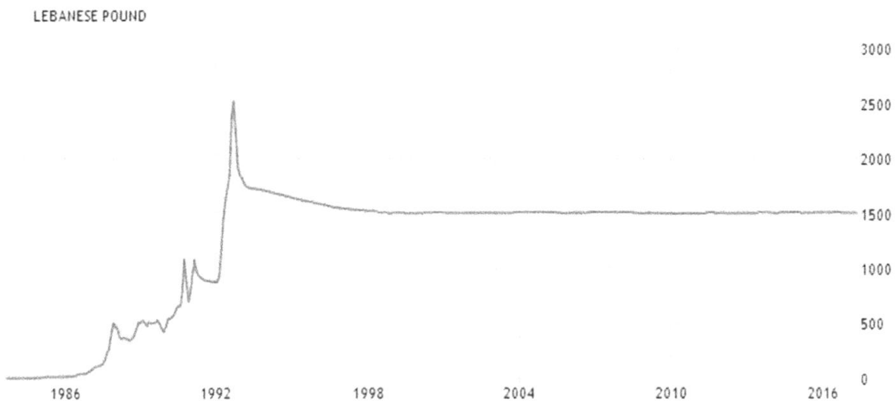

- Switzerland and its most famous banks orchestrated the deportation and dilapidation of the victims of the Second World War (Rickman, 1999). This collaboration with the Nazi German Reich and other dictators raises extreme doubts regarding the Swiss government, mentality, and financial institutions and leads one to conclude they will do anything for financial profit.
- Brutal dictators like the Syria's Bashar al-Assad and his private brigands have a special status in Switzerland, where he owns a more than a billion-dollar bank account (Brown, 2016). Assad has a very strong financial relationship to the Swiss Banque Commerciale Arabe, which was created and managed by the notorious Swiss Nazi banker Francois Genoud, who was supported by Antoine Kamouh and Wafic Said, who financed and organized the Lebanese downslide (Brown, 2016).
- The case and analysis can help governments and people in their daily activities in that they avoid Swiss financial products and services.

The Financial Reason Critical Success Factors

Based on the literature review the most important facts were used as Financial Critical Success Factors which in turn are used in the *Environment*. This section's (CSA: FinancialReason) set of filtered CSFs and their weightings are shown in Table 2.

Table 2. The financial reason critical success factors

Critical Success Factors	KPIs	Weightings
CSF_FinancialReason_Banks	Fall	From 1 to 10. **10 Selected**
CSF_FinancialReason_Tourism	Fall	From 1 to 10. **10 Selected**
CSF_FinancialReason_EconomicGrowth	Altered	From 1 to 10. **10 Selected**
CSF_FinancialReason_CurrencyUsageStrategy	Stopped	From 1 to 10. **10 Selected**
CSF_FinancialReason_FinancialReplacement	Done	From 1 to 10. **09 Selected**

THE MODEL'S IMPLEMENTATION

The proposed mathematical model or the AMM is based on the open group's architecture development method, as shown in Figure 16, where it uses its phases. The AMM is a part of the *Environment* that uses services architecture that enables unprecedented agility and business decision making. These facts can be formalized with a mathematical model that needs a nomenclature.

The mathematical model, as shown in Figures 17 and 18, is presented in a simplified form to be easily readable, bearing the weight of mathematical rigor. The symbol $\sum$ indicates the summation of all the relevant elements, while precise cardinality is omitted.

Each atomic service can represent an (e)neuron that respects the AMM and the AMM is a combination of an organization architecture methodology like the Open group's methodology and a decision system.The AMM is based on a concurrent and synchronized infrastructure, using threads that make various models run in parallel and exchange dynamically data.

The Proof of Concept

The Proof of Concept (PoC) was built using the *Environment* that has beendeveloped using the Microsoft Visual Studio .NET development environment. The PoC is based on the CSFs' indicators binding to a specific research requirement,where the AMM was prototyped using an enterprise architecture tool. The AMM uses atomic services ((e)neurons) that make calls to the AMM microartefacts. The reasoning model represents the relations between this research's requirements, microartefacts or building block and the CSFs.

Figure 16. The Open group's architecture development method (The Open Group, 2011)

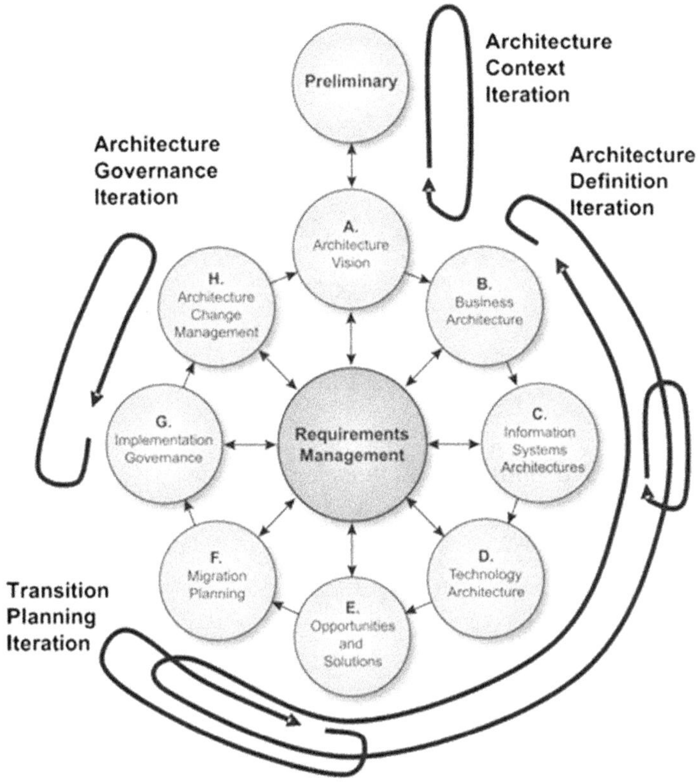

SOLUTION AND RECOMMENDATIONS

The 1975 war was an externally imposed phenomenon that ruined the Lebanon. Invested money was diverted to Switzerland, which has a financial sector very similar to Petro Trad's Lebanon based on private banking. The GDP figures for Switzerland and for the Lebanon show a stunning antisymmetric evolution (Trading Economics, 2017a; Trading Economics, 2017g). This chapter confirms that the bloody and destructive 15-year-long civil war is qualified as a proxy war on the Lebanese soil that caused also damages to France (Hopquin, 2013). Antoine Francis's (2002) works prove that the Lebanese civil war was orchestrated by foreign powers who had precise financial goals.

Figure 17. The model's nomenclature

Model's nomenclature

mcEnterprise	A micro enterprise component
mcRequirement	A micro requirement
mcArtefact	A microartefact
action (or action)	An atomic service (or neuron) execution scheme
mcIntelligenceArtefact	A set that contains: dynamic basic intelligence + governance + persistence+ traceability + data_xsd + resources
mcArtefactDecisionMaking	A microartefact_decision making entity
mcArtefactScenario	A microartefact scenario

Figure 18. The applied mathematical model

AMM

mcRequirement	= KPI	(1)
CSF	= ΣKPI	(2)
CSA	= ΣCSF	(3)
Requirement	= Σ mcRequirement	(4)
(e)neuron	= action + mcIntelligenceArtefact	(5)
mcArtefact	= Σ (e)neurons	(6)
mcEnterprise	= Σ mcArtefact	(7)
(e)Enterprise	= Σ mcEnterprise	(8)
mcArtefactScenario	= Σ mcArtefactDecisionMaking	(9)
IntelligenceComponent	= Σ mcArtefactScenario	(10)
OrganisationalIntelligence	= Σ IntelligenceComponent	(11)
AMM	= ADM + OrganisationalIntelligence	(12)

FUTURE RESEARCH DIRECTIONS

The *Environment* future research will focus on the Lebanese crisis in the period between the year 1975 and 1977.

CONCLUSION

The Lebanese civil war has given no winners and no losers between warring parties, which is unprecedented for a civil war, but in looking at the GDP figures of the Lebanon and Switzerland, it is clear that Petro Trad's Lebanon is a big loser and Switzerland is the unique and immense winner, as shown in Figures 12 and 13. Switzerland's cruel financial behaviour and misdeeds confirm this fact; added to that Switzerland ruling party, the Schweizer Volks Partie, is a racist and anti-Semitic entity that has a 1933-like behaviour (Maurisse, 2016; Miller, 2017; Tribune de Genève, 2014).Ultimately, existing worldwide order cannot prevent such a scenario, which has occurred in many other countries in addition to the Lebanon (Clarke &Tigue, 1975; Abu-Fadil, 2017). The evolution of ethics in finance might bring an end to such financial manipulations and eventually bring to trial countries and their financial system for committing war crimes that caused the deaths of hundreds thousands of people. The Swiss prosecutor Carla Del Ponte a Swiss prosecutor, who was also the Chief prosecutor for war crimes committed in ex-Yugoslavia, hinted generals for alleged crimes (Del Ponte & Sudetic, 2009). The astonishing fact is that why Carla Del Ponte did not prosecute Swiss financial circles who financed these war crimes and arms smuggling in these regions (Hamel, 2003).

ACKNOWLEDGMENT

In a work as large as this research project, technical, typographical, grammatical, or other kinds of errors are bound to be present. Ultimately, all mistakes are the author's responsibility. Nevertheless, the author encourages feedback from readers identifying errors in addition to comments on the work in general. It was our great pleasure to prepare this work. Now our greater hopes are for readers to receive some small measure of that pleasure. This article is dedicated to the Lebanon who has fallen because of greed of financial Swiss giants.

REFERENCES

Abu-Fadil, M. (2017). *Was Intra Bank's Fall a Prelude to Lebanon's Downfall? Huffington Post.*

Aker, F. (2014). *October 1973 - The Arab Israeli War* (1st ed.). Design Pub.

AMInfo. (2014). Middle Eastern clients in the HSBC Switzerland leaks. *Swiss Leaks*. Retrieved June 2017. Retrieved July 10, 2017, from http://ameinfo.com/luxury-lifestyle/list-middle-eastern-clients-in-the-hsbc-switzerland-leaks/

Bitar, K., & de Fouchécour, C. (2015). *Le Cèdre et le Chêne. De Gaulle et le Liban – Les Libanais et de Gaulle*. Editions Geuthner. Retrieved May 3, 2016, from http://www.iris-france.org/publications/le-cedre-et-le-chene-de-gaulle-et-le-liban-les-libanais-et-de-gaulle/

Brown, J. (2016). Wafic Said: businessman, philanthropist and political fixer. *Financial Times*. Retrieved May 15, 2017, from https://www.ft.com/content/a3cb764a-ecf1-11e5-bb79-2303682345c8

Blake, G., & Wagstaff, M. (2016). *The Middle East: A Geographical Study* (2nd ed.). London, UK: RoutLedge.

Caquot. (2017). Sémites. *EncyclopædiaUniversalis*. Retrieved May 18, 2017, fromhttp://www.universalis.fr/encyclopedie/semites/

The International Consortium of Investigative Journalists. (2015). *Banking Giant HSBC Sheltered Murky Cash Linked to Dictators and Arms Dealers*. Retrieved from https://www.icij.org/project/swiss-leaks/banking-giant-hsbc-sheltered-murky-cash-linked-dictators-and-arms-dealers

Clarke, Th., & Tigue, J. (1975). Dirty money: Swiss banks, the Mafia, money laundering, and white collar crime. Simon and Schuster.

Daellenbach, H., McNickle, D., & Dye, Sh. (2012). *Management Science - Decision-making through systems thinking* (2nd ed.). Plagrave Macmillian.

D'Amato, G. (1995). Switzerland: A Multicultural Country without Multicultural Policies? In *The Multiculturalism Backlash European Discourses, Policies and Practices*. Swiss Forum for Migration and Population Studies, Neuchatel, Switzerland.

Dahl, B. (2006). *The LebanesePalestinian Conflict in 1973: The Social (de) Construction of Lebanese Sovereignty*. St. Antony's College.

Del Ponte, C., &Sudetic, C. (2009). *La traque, les criminels de guerre et moi: autobiographie*. ÉditionsHéloïsed'Ormesson.

El Hashem, B. (1990). *It was Kissinger who destroyed the nation of Lebanon*. EIR Feature.

El-Khazen, F. (2000). *The breakdown of the state in Lebanon, 1967-1976*. Harvard University Press.

Faour, M. (2007). *Religion, demography, and politics in Lebanon*. Taylor & Francis.

Favier. (2017). *HaririRafic - (1944-2005)*. Encyclopædia Universalis. Retrieved June 10, 2017, from http://www.universalis.fr/encyclopedie/rafic-hariri/

Fersan, E. (2010). *Syro-Lebanese Migration (1880-Present): "Push" and "Pull" Factors*. Washington, DC: The Middle East Institute.

Fregonese, S. (2009). The urbicide of Beirut? Geopolitics and the built environment in the Lebanese civil war (1975–1976). *Political Geography*, *28*(5), 309–318. doi:10.1016/j.polgeo.2009.07.005

Forbes. (2016). *Edgar de Picciotto& family*. Retrieved July 6, 2017, from https://www.forbes.com/profile/edgar-de-picciotto/

Francis, A. (2002). *Riwa'iat*. Edition Nobilis. Liban.

Habib, C. (2009). *Lebanese Politics and the Tyrany of Confessionalism*. Confluences Méditerranée. Retrieved June 11, 2017, from https://www.cairn.info/revue-confluences-mediterranee-2009-3-page-63.htm

Haley, E., & Snider, L. (2006). *Lebanon in Crisis: Participants and Issues*. Syracuse University Press.

Investopedia. (2017a). *Real Economic Growth Rate*. Investopedia. Retrieved March 22, 2017, fromhttp://www.investopedia.com/terms/r/realeconomicrate.asp

Hamel, I. (2003). Victoire à l'arraché d'un trafiquant d'armes. *Swissinfo*. Retrieved from https://www.swissinfo.ch/fre/victoire-%C3%A0-l-arrach%C3%A9-d-un-trafiquant-d-armes/3212952

Hijazi, I. (1975). *Kissinger Has Nothing to Offer Assad, and Assad Has Nothing for Him*. The New York Times Archives. Retrieved January 13, 2016, from http://www.nytimes.com/1975/03/09/archives/kissinger-has-nothing-to-offer-assad-and-assad-has-nothing-for-him.html

Hitti, P. (2002). *History of the Arabs*. Palgrave. doi:10.1007/978-1-137-03982-8

Johnson, S. (1983). Francois Genoud: Terrorist controller for Swiss banks. *Executive Intelligence Review*.

Hopquin, B. (2013). Attentat du Drakkar: qui a tué les paras français de Beyrouth en 1983? *Le Monde*. Retrieved August 7, 2017, from http://www.lemonde.fr/societe/article/2013/10/23/qui-a-tue-les-paras-francais-de-beyrouth-en-1983_3501317_3224.html#x1dP3v1VIc2M9yL6.99

Indexmundi. (2018). Switzerland vs. Lebanon. *Indexmundi*. Retrieved February 13, 2018, from https://www.indexmundi.com/factbook/compare/switzerland.lebanon/economy

Julian, H. (2014). *Lebanon Jews Denounce Israel as Beirut Synagogue Reopens*. Jewish Press. Retrieved July 10, 2017, from http://www.jewishpress.com/news/breaking-news/lebanon-jews-denounce-israel-as-beirut-synagogue-reopens/2014/05/20/

Kubursi, A. (1999). Reconstructing the Economy of Lebanon. Arab Studies Quarterly, 21(1).

Lebanese Army. (2017). *Jean Njeim*. Lebanese Army. Retrieved May14, 2017, fromhttps://www.lebarmy.gov.lb/en/content/jean-njeim

Le Parisien. (2010). *History of Lebanon under Arab rule*. Le Parisien.

L'Orient le Jour. (2017). Les juifs du Liban partent dans le sillage des guerres de 1967 et de 1975. *L'Orient le Jour*. Lebanon. Retrieved June 11, 2017, from https://www.lorientlejour.com/article/1066188/les-juifs-du-liban-partent-dans-le-sillage-des-guerres-de-1967-et-de-1975.html

Maurisse, M. (2016). *L'enfer des expatriés français en Suisse: une «enquête»*. Suisse: Le Temps.

Miller, J. (2017). Swiss high court rules anti-immigration SVP ad broke racism laws. *Reuters*. Retrieved June 11, 2017, from http://www.reuters.com/article/us-swiss-racism-svp-idUSKBN17F1UT

Miller, M. (2014). *Momentary Memorials: Political Posters of the Lebanese Civil War*. University of Colorado.

Little, D. (2007). *His Finest Hour? Eisenhower, Lebanon, and the 1958 Middle East Crisis*. Oxford, UK: University Press.

Parker, S. (2016). *The demise of secret bank accounts – Switzerland's private banks. face a new era of transparency*. Wolters Kluwer Financial Services, Inc.

Peterson, S. (2011). *Why it Worked: Critical Success Factors of a Financial Reform Project in Africa*. Faculty Research Working Paper Series. Harvard Kennedy School.

Rickman, R. (1999). *Swiss Banks and Jewish Souls by Gregg J. Rickman*. Central European History. JSTOR.

Rolland, J. (2004). Lebanon: Current Issues and Background. NOVA.

Romeo, L. (2011). *Charles Malek. Les Cles du Moyen Orient*. Retrieved August18, 2017, from http://www.lesclesdumoyenorient.com/Charles-Habib-Malek.html

Solomon, Ch. (2017). *The Ghost of AntounSaadeh Looms Large Over Syria*. WorldPolitics. Retrieved June 17, 2017, from https://raddingtonreport.com/ghost-antoun-saadeh-looms-large-syria/

Salibi, K. S. (1976). *Cross Roads to civil war: Lebanon, 1958-1976*. Delmar, NY: Caravan Books.

Sorby, K. (2012). *Lebanon and the 1969 Cairo Agreement*. Archive. ProQuest.

Tabar, P. (2010). *The Lebanese Jewish Community: Emigration and Diasporic Relations*. Institute for Migration Studies and Associate Professor of Sociology/Anthropology - Lebanese American University. Retrieved June 17, 2017, from http://www.ulcm.org/wlcu-lebanese-heritage/news/lebanese-emigration/2013/07/08/the-lebanese-jewish-community-emigration-and-diasporic-relations(dr-paul-tabar)

Taleb, N. (2012). *Antifragile: things that gain from disorder*. Library of Congress.

Tarazi-Fawaz, L. (2014). *A Land of Aching Hearts: The Middle East in the Great War*. Harvard University.

The FactBook CIA. (2018a). *Lebanon*. Retrieved February 13, 2018, from https://www.cia.gov/library/publications/the-world-factbook/geos/le.html

The FactBook CIA. (2018b). *Switzerland*. Retrieved February 13, 2018, from https://www.cia.gov/library/publications/the-world-factbook/geos/sz.html

Toufic, K. (2004). *A political economy of Lebanon, 1948-2002: the limits of laissez-faire*. Library of Congress Cataloging-in-Publication Data.

Traboulsi, F. (2007). *A history of modern Lebanon*. Pluto Press.

Trad, A., & Kalpić, D. (2015). *The (e)Business Transformation Framework for (e)commerce Architecture-Modeling Projects. In Encyclopedia of E-Commerce Development, Implementation, and Management*. IGI Global.

Trad, A., & Kalpić, D. (2016a). A *Transformation Framework Proposal for Managers in Business Innovation and Business Transformation Projects - The role of transformation managers in organisational engineering*. Chinese American Scholars Association Conference E-Leader, Austria.

Trad, A., & Kalpić, D. (2017a). *An Intelligent Neural Networks Micro Artefact Patterns' Based Organisation architecture Model*. IGI Global. (submitted)

Trad, A., & Kalpić, D. (2017b). *Framework and Enterprise Architecture Framework for Managers in Business Innovation-The role of accounting in automated business environments*. IGI Global. (submitted)

Trad, A., & Kalpić, D. (2017c). A Neural Networks Portable and Agnostic Implementation Environment for Business Transformation Projects- The Basic Structure. In *IEEE Conference on Computational Intelligence*. IEEE. 10.1109/CIVEMSA.2017.7995318

Trad, A., & Kalpić, D. (2017d). A Neural Networks Portable and Agnostic Implementation Environment for Business Transformation Projects- The Framework. In *IEEE Conference on Computational Intelligence*. IEEE. 10.1109/CIVEMSA.2017.7995319

Trading Economics. (2017a). *Lebanon - GDP Annual Growth Rate*. Trading Economics. Retrieved April 10, 2017, from http://www.tradingeconomics.com/lebanon/gdp-growth-annual

Trading Economics. (2017b). *Syria - GDP Annual Growth Rate*. Trading Economics. April 10, 2017, from http://www.tradingeconomics.com/syria/gdp-growth-annual

Trading Economics. (2017c). *Israel - GDP Annual Growth Rate*. Trading Economics. April 10, 2017, from http://www.tradingeconomics.com/israel/gdp-growth-annual

Trading Economics. (2017d). *France - GDP Annual Growth Rate*. Trading Economics. April 10, 2017, from http://www.tradingeconomics.com/france/gdp-growth-annual

Trading Economics. (2017e). *United States - GDP Annual Growth Rate*. Trading Economics. April 10, 2017, from http://www.tradingeconomics.com/united-states/gdp-growth-annual

Trading Economics. (2017f). *United Kingdom - GDP Annual Growth Rate*. Trading Economics. April 10, 2017, from http://www.tradingeconomics.com/united-kingdom/gdp-growth-annual

Trading Economics. (2017g). *Switzerland - GDP Annual Growth Rate*. Trading Economics. April 10, 2017, from http://www.tradingeconomics.com/switzerland/gdp-growth-annual

Trading Economics. (2017h). *Cyprus - GDP Annual Growth Rate*. Trading Economics. April 10, 2017, from http://www.tradingeconomics.com/cyprus/gdp-growth-annual

Trading Economics. (2017i). *Switzerland's currency evolution*. Trading Economics.

Trading Economics. (2017j). *Lebanon's currency evolution*. Trading Economics.

Tribune de Geneve. (2014). *Pour Christian Levrat, l'UDC est sur la voie du fascisme*. Tribune de Geneve. Retrieved July 12, 2017, from https://mobile2.tdg.ch/articles/551c50ecab5c3729aa000002

UN. (n.d.). *Taif Accords*. UN. Retrieved April 10, 2017, from http://peacemaker.un.org/lebanon-taifaccords89

Vloeberghs, W. (2015). *Architecture, Power and Religion in Lebanon: Rafiq Hariri and the Politics of Sacred Space in Beirut*. BRILL.

World Journals. (2002). *Petro Trad*. World Heritage Encyclopedia. Retrieved from http://worldjournals.org/articles/Petro_Trad

Zalloua, P. (2004). *The NGM Study "Who were the Phoenicians" and the Return of the Phoenicians*. Interview Lebanese Broadcasting Corporation (LBC). Retrieved April 16, 2017, from http://phoenicia.org/genetics.html

KEY TERMS AND DEFINITIONS

Business Engineering: Is the development and implementation of business solutions using business models to business processes and organizational structure.

CSFs: Can be used to manage the statuses and gaps in various project plans and gives the projects the capacity to proactively and automatically recognize erroneous building blocks and to just-in-time reschedule the project plan(s).

Enterprise Architecture: Is a known defined practice for driving enterprise analysis, design, planning, and implementation, using a holistic approach with the use of a system's strategy.

Environment: Is this research's framework.

Manager: Business Transformation Manager.

Project: Business Transformation Project.

Strategy: Is a high-level plan to reach one or more objectives under defined constraints.

Chapter 5
Movie Making in Conflict Zones of Nigeria:
The Jos Experience

Edward Ossai
Plateau State University, Nigeria

ABSTRACT

Filmmaking in Nigeria has no doubt attracted so much attention worldwide and the industry has become a major employer of labor and a source of national revenue. It has also given birth to regional language and ethnic film industries, including Kannywood and Yoruwood. These industries have become a voice for the voiceless, a medium to express cultures, values, and socio-political issues. However, the process of movie making in Nigerian conflict areas has been marred by seemingly sustained ethno-religious conflicts and attacks by insurgents in major film release points like Jos, Lagos, Kano, Kaduna, Aba, Onitsha, and Abuja. This discourse is informed by the need to interrogate how conflicts are detrimental to collective development of the Nigerian state and the film industry in particular. The Jos experience has generated concerns in the industry, hence the need to highlight reason(s) for the drop in film production on the plateau.

DOI: 10.4018/978-1-5225-5987-0.ch005

INTRODUCTION

The art of movie making in Nigeria has improved the quality of Nigerian movies and added value to the entertainment industry generally, thereby reducing the craving for foreign movies. The film medium has intervened in shaping the society in all ramifications. This is in addition to opening windows of opportunities for the screenplay writer, director, cinematographer, camera operator, editor, set designers and all other craft men involved in the business of movie making. These professionals can make the audience believe every picture and sound portrayed in the movie, principally because movies have the capacity to transport audience into many worlds at the same time.

Movie making requires discipline, concentration, a good dose of creativity and patience. It is better to make movies in enabling environments for effective and efficient performance and output. Principally speaking, movie making requires that the movie maker creates his story, hires his crew and cast, scouts for location, dress the set, source, gather and test equipment, make the movie, edit, add sound effects and music, create the title and credits sequences, export the movie to any format and release. However, the security of the crew, cast and equipment are paramount. The location, crew and cast are considered very crucial in making movies in conflict zones. Locations generally refer to the acting arena, which could be an office, room, toilet, football field, Forest and so on. These locations could be used at night or during the day and the locations are determined by the script.

In addition to the space required for the film shoot, the source of light (windows, doors or electricity] is also important since it determines the colour quality of the movie. What the crew does for the movie maker is to help him translate the script into pictures and sound. This also implies that the crew and cast ought to operate in an enabling environment to be creative and efficient. Creaativity cannot be maximized in an atmosphere that is largely violent and chaotic. How can economic prosperity, intergration and peaceful co-existance thrive in hostile settings. The cast and crew may not work together and be committed if they hold strong ethnic and religious views against one another. The movement of crew, cast and equipment therefore becomes difficult. Major film and video film release points like Jos, Kano, Kaduna, Lagos, Onitsha-Aba, Enugu, Calabar, Abuja and Portharcourt have experienced one form of conflict or another and this has affected film making generally. Nigeria has experienced a number of violent conflicts after the 1967-1970 civil war. There have been conflicts between Modakeke and Ife, Aguleri and Amuleri, Itsekeri, Ijaw, Urhobo and Isoko, insurgency in north east and central Nigeria. Selected cases of Ethno-Religious Conflicts, attacks by Fulani Herdsmen and unknown gunmen in

Nigeria indicate that Plateau State has been a flash point between September 7th 2001 and 13th March 2018. For instance, the September 7th, 2001 conflict was between Hausa\Muslims and Christian\ Indigenes. The subject matter was the political appointment of a non-Plateau in Jos North, a position the Plateau Youth Council believe belongs to them. Lives and properties were lost as a result of this conflict. Bagudu N. (2004),reveals that the BarkinLadi crises of December, 30th 2001, the Jos mayhem of May 2rd 2002, the Bassa ethnic clash with the Fulanion May 27th, 2002, the Yelwan-Shendam and Wase conflicts of June and July, 2002, spilled into Langtang North and Kanam in March 2003. According to Matawal (2012):

The violent crises that followed in 2002 to 2004 were mostly in the nature of reprisals. In fact, 2004 was the turn of the Southern Senatorial District of the State. The crises overwhelmed five of the six LGAs of the Southern zone and Kanam in the central zone too, was affected. The worst hit was Yelwa in Shendam LGA. (p102).

Attacks, reprisal attacks and counter attacks continued in Quanpam, Wase, Shendam, Jos-North, Jos-South, Maingo, Riyom,BarkinLadi and Dogo Na Hawa on the 8th March 2010. The conflicts are still on going and the open show of hate and animocity against one group or the other could be one of the reasons for the sparks in major flash points on the Plateau. The State is so volatile that any little argument could lead to a major crises resulting in the loss of lives and properties. As recently as Monday 26th February 2018 a dispute between officials of the State government and tricycleoperators over an increase in tariff lead to a serious commotion.Businesses were shut down as everyone scampered for safety. Even more recent is the Monday 12\13 March 2018 killings and wanton destruction of houses and farmlands at Dundu, a village in Kwall District of Bassa, Plateau State. Sahara Reporters, an online news agency, reports tha tTwenty three (23) were killed in fresh attacks in Plateau (2018).Retrieved from http://saharareporters.com/2018/03/13.These and many more disputes have contributed to the drop in number of activities, including film business on the Plateau.

The insurgency in the north eastern Nigeria has led to the destruction of lives and property and it has also rendered millions homeless. A lot of them, though trumatised with no hope of ever recovering have found a haven on the Plateau. This could explain the population explosive on the Plateau and its environs. Two questions now arise: How can all these conflicts be curbed with the rivalry, competition and suspicion among the various ethnic groups in Nigeria? How can movies be produced in an atmosphere that is largely violent and chaotic?

CONCEPTUAL CLARIFICATIONS

This discourse articulates the relationship between moviemaking and conflicts as major variables. The framework of analysis is therefore based on the premise that the environment required for the sustainance of moviemaking is undermined by protracted violent conflicts. Moviemaking components cannot be optimally reached in an atmosphere of violent conflict.

Apart from being the society's watchdog, film and or the maker is a molder of the society, therefore film is not created from a vacuum. Thus, film is ideologically conditioned and made to support varying ideological positions. What this implies is that films can be used to articulate varying ideological interests and to generate discourses that have to do with the people's socio-political and economic development and their social responsibilities. Film studies have been perceived from various view points and they are understood within specific conceptual models of analysis. For the purpose of this discourse, film, movie and video film will be referred to as film or motion picture. That is, moving pictures as against still pictures. Motion picture in this context is a series of still photographs on film projected in rapid succession onto a screen by means of light and because of the optical phenomenon known as persistence of vision; it gives the illusion of actual, smooth, and continuous movement.

Conflict on the other hand is considered as a serious political and economic problem which deters the norm of constructive dialogue, negotiation and integration. Political violence and conflicts are usually fallouts of frustration, poverty, marginalization, bad governance and extreme levels of inequality. This seem to agree with Frustration theorists who argue that frustration can lead to dissatisfaction, disaffection and then to conflicts. Many scholars have contributed to propounding the theory of Frustration-Aggression in their effort to explain why violent conflicts are so predominant in the human society. Scholars like Freud (1939), Dollard (1939), McDougal (1960) and others have contended that frustration most often leads to aggression and war. They submit that once people are dissatisfied with a given situation, they become frustrated. If for any reason they do not get out of such situation, they tend to promote certain tendencies within them that are supposed to redress the social injustice, but in the process, such situations lead to aggression and war. (Dougherty & Pfaltgraff, 1997). This suggests that conflict is part of man and the society and it is a reality of life. This also indicates that when disputes are not quickly and logically handled serious conflicts are bound to happen. The eruption of conflicts and ethnic wars in Europe, Africa and Asia punctuated the fact that this is not a problem specific to the developing world but a security threat to developed states as well. Conflicts are of different magnitude for different reasons. Some of them are economic, political, ethnic, while others are religious or land disputes. Shedrack Gaya Best (2008) posits that causes of conflicts are many and varied: These are distinguishable from the triggers of the conflicts in the respective flashpoints. By causes we mean those deep-seated factors that lead to conflict and must be addressed and transformed to achieve sustainable peace. (p17).

Some of the conflicts were resolved through round table discussions or outright confrontations, while others are ongoing. The conflicts in Jos and its environs fall under the later. And because it is ongoing some investors have relocated, while others are operating on a low key.

AN OVERVIEW OF FILMMAKING IN A SELECTED CONFLICT ZONES

A study of Film development in Europe, America, Asia and Africa reveals that it started at different periods under peculiar situations. The politics, economics and sociology of the periods dictated the pace at which it developed. According to film scholars like Whetmore (1985)

Film production began at about the same time as the popular illustrated press, gramophone, radio, television and other technologies which came into existence only during the nineteenth century. What this translates to is that modern film production can be found in the work of inventors, engineers, scientists and craftsmen, who introduced diverse social values. Documented history reveals that scientists of British, French, German and other European origins conducted series of independent and sometimes interrelated experiments on film strip, light, camera, projector, sound and so on. The results from one experiment led to other experiments (p 204)

Countries like United States of America, United Kingdom, Germany, France, Russia, India, China, South Africa, Kenya, Senegal, Burkina Faso, Italy and Nigeria are considered hubs of movie making in the world. They have all experienced one form of conflict or the otherthat has affected film production since the inception of film in 1897. They are film production hubs as well as conflict zones. From the first and second World Wars to the attacks and activities of Islamic extremist in Orlando, USA, Nice in France, Nairobi Kenya, Italy, Turkey, Afghanistan, Israel, Syria and Nigerian cities have affected economic activities and particularly in the area of film production in Nigeria.

Some communities in Nigeria have used various strategies to settle disputes and conflicts. Communities like Ajegunle in Lagos, Warri Communities in the Niger-Delta, Shagamu in Ogun, Jukun-Tiv, Toto, Obi, Nasarawa-Eggonin Nasarawa State, Jos-Yelwan -Shendam, Wase and Langtang in Plateau State, Kano, Kaduna and the whole of North East, Nigeria. Conflicts, be it ethno-religious or political have devastating consequences. Conflict situations could be so dismal and insecure for citizens to dwell not only where they desire, but even where they once claimed to be their place of descent because of perpetual fear, suspicious, perceived threats and insecurity. Movies are better made in enabling atmospheres for effective and efficient performance and output.

MOVIE MAKING: THE JOS EXPERIENCE

The Jos experience indicates a situation where security operatives are strategically positioned at flash points and major roads. Vehicular checks are regular sights as vehicles, drivers, as well as passengers are subjected to frustrating screenings. People move around in fear and in anticipation of being molested or attacked by unknown persons or groups. This implies that a movie maker in a conflict zone cannot specifically determine or predict when he/she will start and wrap uphis/her film production.

This discourse is not to suggest that Jos is a war-torn zone, but to share the fact that investigations by the researcher, who is also a participant observer in the political developments of Jos that some business men and women have found Jos a haven for business. Some other Nigerians relocated to Jos to acquire knowledge in one field of endevour or the other. Jos, also referred to as the Tin City, is a maniture Nigeria.

Lar, Isaac and Reuben Embu (2012) reveal that:

Plateau people are peace-loving, kind and generous, receiving and hosting people from other parts of the nation. This serene world of mutual co-habitation, melting pot of national cultures, was rudely dislocated by the ethno-religious violence which broke out on the 7th september 2001. This show of hostility was a climatic version of earlier sporadic clashes bordering on social, economic and political issues [p xi]

The position of Lar and Embumay explain why commercial activities have dropped to a very low eeb, thereby affecting the economy of the zone and the production of movies. Film business in Jos dates back to 1952, with the first documented public film screening at Vwang, a town near Jos. Subsequent screenings were held at Yelwa club, Bukuru and Plateau clubs, years later. When the British colonialists moved into Jos, Plateau for tin mining, they produced mainly documentary films to educate, entertain and enlighten the natives on their activities and on health related issues. They moved from one village to another in a Land Rover vehicle screening films at village squares and market places. Dung Pam Sha (2005) notes that: The establishment of colonial Tin Industry in the 1880s led to a significant expansion of migration into tin mines on the Plateau[p 34].

What resulted from this migration was the urbanization of Jos city. The fine climatic condition, beautiful landscapes, sceneries and the establishment of the first colour Television in West Africa in Jos and the Jos campus of the University of Ibadan Theatre Arts department in 1974 and 1976 respectively stimulated the discovery of many talents in various aspects of filmmaking and entertainment generally. The presence of the Theatre and Film Arts department of the University of Jos, Nigerian Film Corporation, NTA Television College and the National Film Institute further exposed film potentials in the Jos Plateau. .

Subsequently, Jos became the home of many film and television stars, scholars, producers and sports men and women. Some of them include: George Menta, Peter Igho, Chris Orakpotobo [Shanana] Ene Onoja, Lantana Ahmed, Zuberu Attah, Saint Obi, Franca Brown, Eddy Fom, Frankie Ogah, Sani Mu'azu, Suleiman Adara, Ali Nuhu, Ireti Doyle, Rakhiya Attah, Steve Gukas, Akume Akume, Yakubu Lamai, Kasimu Yaro, Sunday Bulus Rogo, Matt Dadzie, Ladi Totti, Sadiq Daba, Soni Oti, Fred and Zack Amata, Izu Ojukwu, Kenneth Gyang, Steve Akasa, Zainab Bewell, Gladys Dadie[Bilkisu Gubadia], Suru Mentie, Hussaini Baba Ahmed, Yusuf Mohammed, Magaji Mijinyawa, Samuel Ajayi, John Ndanusa, McAuthur Fom[Nosa], Abraham Doe, Emma Oniwon, Ignis Ekwe, USA Galadima, Brendan Shehu, Chika Okpara[Zebudaya], Maureen Egbuna, Dalhatu Bawa, and many others.

These individuals no doubt contributed immensely one way or the other in making Jos, Plateau the hub of film and entertainment business in the Middle Belt zone of Nigeria. This is evident in the number of films shot and broadcast in Jos and nationwide. Some of them include, Soweto, Mr. Johnson, Cock Crow at Down, Behind the Cloud, Supple Blues, Shadows, Riddles and Hope and many home videos. Cock Crow At Down for instance, was shot in Jos and broardcast nation wide for about ten years. Nigerian filmmakers like Izu Ojukwu, Teco Benson, Okey Ogunjaifor and a few Lagos based filmmakers prefer to shoot their films in Jos because of the serenity and cost of production. They claim that Jos is not as congested or as chaotic as Lagos.

However, it is imperative to note that the number of movies produced in Jos, Plateau has dropped drastically for the very obvious reason of insecurity. It is imperative to note that as a result of the conflicts and the risk involved in making movies in Jos, many movie makers have relocated to more enabling environments. At the heat of the crises it is became common to see people running for safety whenever they hear the shout of "Anfara'' in Hausa, meaning, ''They have started''.

In a personal interview with SaniMua'zu, a resident filmmaker and the Executive Director, Lenscope Media, Jos, laments that:

It was really sad that I had to relocate to Kano with my family for a while. My house and office were attacked so I had to run for my life. I cannot explain what really happened but it was devastating. I had very talented Christians and Muslims on my staff list who related as brothers and sisters. Nobody bothered about where you came from. Thank God I am alive and back to Jos, but it is not business as usual.

Mua'zu expresses serious disappointment in the system that does not allow him do business in the town of his choice, a town he has lived all his life. In a personal interview with him, he suggests the review of the Nigerian constitution to enable Nigerians, no matter their callings, ethnic or religious affiliation live where ever they want.Ali Nuhu, a Nigerian film actor and Director also relocated to Kano for obvious reasons. Apart from the insecurity in Jos, he claims that his movement to Kano was also economic because there is more business in Kano compared to Jos. In a personal interview with him during the film shoot of *Queen Amina* in Jos, he states that ''Jos is like home to me, but I decided to set up my business in Kano because of its commercial and metropolitan nature. Jos is no longer the Jos of old because of the insecurity of live and property''.

However, there are some people who could not relocate for specifics reasons which has to do with the fact that they have already established themselves with lots of investments in Jos. Some filmmakers from other parts of the country still prefer to shoot in Jos principally because of the story they plan to shoot. Such people are condemned to make their movies under lamentable conditions. So far, not all of them have had the same experience.

The movie, *Zanga Hill* [2012], was shot in Jos at the heat of ethno-religious violent killings. Oneof the lead actors, Edwin Hayes described the experience as nightmarish. In a personal interview with him, he states that:

Principal photography started on the 21st of May 2012 at 9am exactly. Initial preparations included permissions from the Police and the Special Task Force, code named Operation Safe Haven. The location was Shere Hills, located on the outskirt of Jos. For this film to be shot, the cast and crew had to be camped at the foot of the hills. Welfare was prepared by some cast members and we had a cast and crew of seventy people. We spent a total of one month, two weeks.

Many other movies have been shot after this. The most recent one is *Queen Amina*, directed by Izu Ojukwu and produced by Okey Ogunjiefor. It was shot in the same area., that is Shere Hills area in Jos East. Though some scenes where shot at the MOTNA, around the musuem, most of the war scenes were short in Jos East, indicating that it is a safe haven.

However, in a personal interview with one of the assistant directors of the movie, Sidi Ishaqreveals that-

First of all, it was during the last phase of our research when the producer came that we assessed the security issues. I convinced him the more during our auditions in Kaduna and Zaria, evidenced by the large turnout. That turnout gave my producer hope. Later we discovered our locations proximity was not convinient so we thought of scouting Jos where we found the befitting locations. It was when we settled to finally shoot in Jos that we wrote officially to security agencies... using mitigation measures, locations were also chosen considerably close to security reach... Our camp was close to the Army barracks, Command Centre of the Joint Task Force{ JTF}, which is also close to our location.

There is no gainsaying that it took a lot of convincing and persuasion to get the Lagos cast and crew to come to Jos for the shoot. The reason is obviously the security situation in the state. Three major crises that led to the drop in film production in Jos are the April 12, 1994, April 10, 1997 and September 7, 2001 ethno-religious conflicts. These crises were followed later by pockets of selected attacks in 2002, 2004, 28th November 2008, 17th January 2010 through 2015 by ' Unknown gunmen', ' Fulani herdsmen', hoodlums and members of the Boko Haram.

According to Dung Pam Sha(2005):

The 1994 and 2001 conflicts were conflicts involving Hausa\ Fulani versus all the non-Hausa\Fulani and christians thus taking a dimention of a national conflict in the city of Jos...The Nigerian State played a direct or indirect role in the creation of these conflicts in an attempt to appease one section of the communities in Jos. All the conflicts are rooted in the crisis of citinship. The historic divide between civic and ethnic citinship has been responsible for these conflicts. [pp74-75]

One can make bold to say that the struggle to acquire land forcefully by some settlers is responsible for some of the contestations on the Plateau. The government's position on the Gyero crises is that:

Government should make it abundantly clear that the issue of indigenisation is a constitutional matter and not just a matter of claim by any group no matter how long they have stayed in the area. It is a national issue. This is not peculiar to Plateau State. This is without prejudice to the freedom of any Nigerian to live anywhere and own property according to law; and pursue his livelihood legitimately. Those who have come to settle in a particular place should be made to realise their status as such and no more. In the case of Jos, it is wondered what would have been the situation if other ethnic groups from other parts of the country who have settled here for a long period as Hausa\ Fulani have also been laying claims to the indigenship of the area [p 16].

In his well research work on 'Ethnic Conflicts and Citizinship crises in the Central Nigeria' Alubo (2006) contends that:

Nigeria's plural composition of over 370 ethnic groups is well known . Less recognised are the challenges this diverse composition poses to national integration, a sense of belonging and participation in decision-making. This lack of recognition is manifest, among others, in the jettisoning of the independence national anthem, which recognised that 'though tribe and tongue differ, in brotherhood we stand' and its replacement with the bland ' arise o compatriot, Nigeria's call, obey'. However just who the compatriots are is contested as some Nigerians are excluded from participation in activities, or benefitting from the same opportunities as are available to other 'compatriots'... The issues arecomplex and serve as reminders that questions of identity and citiinship are **too** *crucial for any attempt to forge national unity and integration to be wished away[p 1]*

It is therefore obvious that Nigeria is not the only plural state in Africa or the world. Countries like Malaysia, Brasil, India, Pakistan,Canada, America and so on have evolved a workable system to contain their plurality. This simply indicate that there is strength in diversity. Matawal (2012) remarks that:

The indigene-settler issue is a deep-seated matter in the national question which is not peculiar to Plateau State. This phenomenon is seeking to undermine the tenets of Nigeria's federalism; thus reminding us of the mistake of Lugard's 1914 amalgamation of Nigeria [p 104].

The Nigeria situation is such that the Army is not committed to the fight against Boko Haram, especially with the recent kidnap of students of Government Science School, Dapchi, The former Minister of Defence, Gen. Theophilus Yakubu Danjuma recently accused the Nigerian Army of allegedly colluding with bandits to kill and maim the people of Taraba and others outside the State. Monies meant for troops are ambushed and carted away by top military brass. Monies meant for the purchase of arms and aminations are misappropriated, troops are forced to flee battle fields because they are ill equipped and government does not seem to have unity of purpose. The Nigeria Army has been fighting a group that is more equip than it is and has the capacity to gather intelligence reports faster. The good news however is that the Nigerian President Mohammadu Buhari has declared that 'technically, the Nigerian Army has won the war against Boko Haram'. This is debateable considering the fact that the insurgents are still in control of some Nigerian territoriesand their victims are mainly armless civilians, many of them women and children.

Again, has film bussiness faired well with the contant distrust, disruption of peace and economic activities in Nigeria and particularly on the Plateau? The National Film Institute, Jos use to produce five short films every semester, a fallout of the Production Workshops. The reseacher was part of these Production Workshops between 1999 and 2013. A period that experienced gradual drop in the number of films produced and the number of facilitators from outside Jos. In other to surmount this challenge, facilitors were sourced from within the Film Institute Alumni, thereby undermining the philosophy of the Production Workshop, which is to expose students to industry professionals. Again, even among the Alumni, some will rather remain in Lagos or any town outside of Jos. The philosophy and practice of bringing industry professionals has truly empowered the students. A lot of them [Ishaku Gomut, Kenneth Gyang, Yinka Edwards, Steve Akasa, Achor Yusuf, Leo informi, Julius Morno, Idegho Kagho, Sidi Ishaq, Izu Ojukwu, David Uro, Ik Okeagu, Fortune Ejim and so on] are making waves in the industry today as camera men, scriptwriters, costume designers make-up designers and or directors. But things are nolonger the same. The industry professionals are not interested or excited about coming to Jos.

CONCLUSION

The filmmaker, also considered the society's watchdog, a commentator, communicator, bridge builder, facilitator and harnesser can use movies to shape the minds of the warring ethno-religious groups positively. An enabling environment for filmmaking therefore becomes paramount for both film development and peaceful co-existance. Principally, the film medium has become a voice for the voiceless, a medium to express cultures, values and socio-political issues. If properly harness it could be used to bridge ethnic and religious diversities in Nigeria. However, it has been difficult dealing with insurgency and militancy because of the nature of the war farer. A machinery for reconcilaition and reconstruction must be put in place in order to prevent and manage conflicts no matter their nature. Government and its law enforcement agencies should not wait until a conflict has developed into violent action and reaction and counterraction and all manners of reprisal attacks before they intervene. The early warning system should be adopted. The judiciary must be empowered to punish all conflict merchants no matter their position in the society. The Nigerian military, whose mandate is to protect the Nigerian State from internal and external aggression must be trained in modern warfair and equipped to contain both conventional and unconventional warfares. With all these things in place, Jos has the potentials of taking its rightfull place as the haven of film production, sports and entertainment generally. Most importantly, business will flourish and socio-economic development will improve and peaceful co-existance will hold sway,

REFERENCES

Matawal, A. B. (2012). The Jos Crises and National Security: A Historical Perspective. In *Creative and Critical Writing for Peace Building in Nigeria: The Jos Paradigm*. Akin Press.

Dougherty, J.T., & Pfaltgraff. (1997). *Contending Theories of International Relation* (4th ed.). Addison Wesley, Education Publishers Inc.

Dung, P. S. (2005). *The Politicisation of Settler-Native Identities and Ethno-Religious Conflicts in Jos, Central Nigeria*. Stirling-Horden Publishers Ltd.

Imobighe, T. (2003). Ethnicity and Ethnic Conflicts in Nigeria: An Overview. In Civil Society and Ethnic Conflicts Management. Spectrum Books Ltd.

Elaigwu, I. J. (1994). *Ethnicity and Federal Option in Africa*. Abuja: National Council on Intergovernmental Relations.

Lar, I., & Embu, E. (2012). *Creative and Critical Writing for Peace Building in Nigeria: The Jos Paradigm*. Akin Press.

Alubo, O. (2006). *Ethnic Conflicts and citizenship crises in the central region*. Ibadan: Programme On Ethnic and Federal Studies.

Umar, M., & Ossai, E. (2014). Ethnicity and The Dilemma of Integration in Nigeria. *Current Issues in Development, 2*(3).

Oteh, P. J. (2008). Our House: A Post-Colonial Performance Of Colonies. In *Crisis, in A Window on Africa: Changing Culture and Peoples*. Lagos, Nigeria: Dat and Partners.

Ugor, P. (2013). Nollywood and Post-colonial Predicaments: Transnationalism, Gender, and Commoditisation of Desire in Glamour Girls. In *Global Nollywood; The Transnational Dimentions of An African Video Film Industry*. Bloomington, IN: Indiana University Press.

Plateau State of Nigeria. (1997). Government Views and Decisions on the Report of the Administrative Committee on Gyero Communal Clashes in Jos South Local Government Area. Ministry of Information and Communication.

Shedrack, G. B. (2008). *Causes and Effects of Conflicts in the Southern Zone of Plateau State Nigeria*. Centre for Conflict Management and Peace Studies, University of Jos.

Whetmore, J. E. (1985). *Media America* (3rd ed.). Wadsworth Publishing Company.

Scores Killed in Fresh Attacks in Plateau. (n.d.). Retrieved from http://saharareporters.com /2018/03/13

The Sun Newspaper. (n.d.). Retrieved from sunnewsonline.com/killing-ty-danjuma-dhq-and-the-poor-people-of-taraba/amp/

Chapter 6
Women, War, and Conflict:
A Thematic Discourse on Femi Osofisan's Women of Owu

Festus O. Idoko
University of Jos, Nigeria

ABSTRACT

Conflict and war are not unique to Nigeria or Africa. It is now a truism that in conflict situations women are one of the worst hit. The reason is not farfetched: women are in conflict areas, and they are likely to be raped, kidnapped, killed, and or end up as widows/concubines. Although women are seldom directly involved in conflict situations, they serve as the first line of aid, providing what can best be described as succor to the casualties of conflicts as well as maintain the home front in the incessant absence of the male and youths. Predicated on critical feminist theory and the relational theory of conflict, the chapter from a thematic view discusses the plight of women in conflict situations both within the text and the context that is Nigeria. Using Femi Osofisan's "Women of Owu" (2006), an adaptation of Euripedes' "The Trojan Women," the chapter argues that the manner in which Osofisan dramatizes the story of the African woman trapped in conflict and war is both sympathetic and resilient, yet a gruesome reminder of the position women find themselves in in conflict situations.

DOI: 10.4018/978-1-5225-5987-0.ch006

INTRODUCTION

Freire's (1970) assertion that oppression is almost always manifested in leaders who shy away from dialogue, and more so where such leaders insist on imposing their decisions. This manipulative approach does not organize the people or liberate them. This sets the tone for this discourse and throws some light on the implications of a dialogically barren leadership, since the lack of constructive dialogue itself can lead to conflict and even war. This assertion also draws attention to the contradictions of dictatorial leadership that employs the weapon of manipulation. Both implications are in effect tantamount to oppression which is a precursor to conflict and war.

In an axiomatic sense, existence here on earth is necessarily conflictual. In fact, in many conflict situations, women are one of the worst hit. The reason this is so is the irreducible fact that life itself is not a perfect bed of roses. Disagreements almost always degenerate into conflict and if this is not carefully managed and resolved it can result in war. This scenario has amply been reflected and mediated in several dramatic texts.

The paper argues that the manner in which Osofisan retells the narrative of the African woman (*ala Womem of Owu*), is both sympathetic and resilient, yet a gruesome reminder of the disadvantageous position women find themselves in conflict situations. In addition, the paper draws attention to the fatal hope of the women in the promised revenge of the gods and argues that Osofisan's subjugation of the role of the gods yet again shows how much less faith we should have in the gods.

The paper concludes that in reality, the irony of liberation is that the triumph of any side, unless based on genuine and fair grounds, would only have set the stage for yet another conflict sooner than later, and the likelihood that women would be worse hit is still very high. This irony, within the literary text under review, is accentuated by the fact that more often than not, literary texts (owing to the nature of their literariness and coupled with the low level of literacy), in Nigeria and Africa at large, are not available to a larger majority of the populace who are supposed to be the vanguard of the liberation, and here, women constitute a larger percentage.

Conflicts in the Dramatic Text

While it is a fact that playwrights draw from the on goings within the society ranging from the social, the religious, cultural, to the political and the historical, the flip side is that what drives the inner actions of the play is conflict and most if not all playwrights seek to resolve this conflict that drives the actions of the play. For instance, Ola Rotimi (1976) in his play *Kurunmi* demonstrated the fact that the conflict over tradition coupled with the refusal of the ruling elite to embrace dialogue led to the war in the play between those who support that the tradition be upheld and those who desired to change it. Similarly Rotimi (1979) in *The Gods are not to Blame* (through metaphoric postulations) made the point that it was the conflict over allegiance to tribe (as exemplified by the protagonist-Odewale) rather than love for country that led Nigeria into her first civil war.

In addition, Irene Isoken Salami-Agunloye (2008) in *Idia, the Warrior Queen,* we see a situation where a conflict between one of the King's Chiefs over the latter's wife eventually blossomed into war when the Chief switched allegiance to the rival Igala Kingdom. Furthermore, in *Emotan* (2001), Emotan's quest for justice motivated her to mobilize popular support for the exiled heir prince leading to the overthrow the King and restoration of order to the Bini kingdom.

In all these instances of conflicts women are usually caught in the cross fire even where they are not actually in the forefront of the war. Women are likely to be killed and or end up as widows. Although women are seldom directly involved in conflict situations, they serve as the first line of aid, providing what can best be described as succor to the casualties of conflicts as well as maintain the home front in the incessant absence of the male and youths. Predicated on critical feminist theory and the relational theory of conflict, the paper, from a thematic view discusses the plight of women in conflict situations both within the text and the context that is Nigeria- using Femi Osofisan's (2006) *Women of Owu* (an adaptation of Euripedes' *Trojan Women*).

THEORETICAL AND CONCEPTUAL FRAMEWORKS: WOMEN AND CONFLICT

Dennis Ityavyar and Stella Obiajunwa (1992) have made the point that;

The Nigerian woman is not just the product of her biological origin or her social structure. She is equally the product of the various state policies that are designed within the context of the dominant male ideology (p.2).

This is a conceptual truism. It implies that the Nigerian woman has been unequally yoked with the male counterpart, trapped by biology, and social stratification within a patriarchal matrix. Ityavyar and Obiajunwa's assertion above is corroborated by Omonubi-McDonnell position that; "the Nigerian women confront two distinct yet related systems of power: the economic mode of capitalism and the ideological mode of patriarchy" (as cited in Ityavyar and Obiajunwa, 1992, p. 2). The two positions become clearer when collocated against the ripples occasioned by colonialism on Nigerian/ African people. This is arguably because;

The African woman was never aware of gender problem in society until the advent of colonialism. Not that gender problems never existed. ... the structural integration of Nigerians into the mainstream capitalism via colonialism has had adverse effects on the roles of women. Colonialism and capitalism brought a new social order. The new socio-economic order was predicated on Victorian Conceptions of womanhood. The African educational, legal and social system was altered in disfavour of the African woman (Ityavyar and Obiajunwa, 1992, p. 3).

In the contemporary African society it appears that any appreciable discussion on women draws us back to gender. By gender, we refer to "the words, gestures, appearances, ideas and behaviours that dominant cultures understand as indices of feminine or masculine identities" (Emenyonu, 2000, p. 29). However, our theoretical starting point is the definition by the World Health Organisation (2018) website that gender refers to:

The socially constructed roles, traits, attitudes, behaviours, values, responsibilities, relative power, status and influence ascribed to male and female humans on a differential basis. Gender identity (masculinity/femininity) is not biological, but learned. It is changeable over time, and varies widely within and across cultures. Gender refers not simply to women and/or men, but to the relationships between and among them. Gender identities condition the way human beings are perceived, and how they are expected to think and act.

This broad definition sets the tone for a holistic conceptualisation of gender. However, the concept of gender is a thorny issue in itself given the manner it has been misapplied in discourses. This explains why Carpenter (2002) asserts that "gender theory thus should distinguish between sex and gender: use both distinctly as independent variables; and seek to analyze the scope, context and logic behind what overlaps exist between the two" (2). Nevertheless, the misapplication of the concept of 'gender' appears to be as a result of the influence of sex-role theory since as Kathleen Trigiani (n.d) puts it on her webpage:

Sex-role theory implicitly assumes that our culture values the characteristics of each sex equally and that they complement each other in a balanced way. Women are just as esteemed for their passivity as men for their aggressiveness. Sex-role advocates will admit these traits have their limitations.

However, they always assure us that in the final analysis, conforming to our roles will bring glad harmony between the 'sexes'. This is a specious argument that leaves much to be desired in the worst case scenario. It is partly in this sense that Carpenter (2002) posit that;

gender is a more complicated concept than sex.: ...Where sex is about maleness or femaleness (measured in biological difference), gender is about masculinity and femininity, measured in socially constructed and contextually contingent ideas about what attitude, attributes and behaviours correspond to ideal-type maleness or femaleness.

The point Carpenter is making is that the 'concept of gender help explains variations across same-sex distribution, culture and time periods.

The concept of sex explains uniformity within same sex distribution and variation between the sexes. As Carpenter (2002) puts it,

Identifying gender as a social construction, fluid and malleable to suit the interest of the powerful, opens a broad explanatory power over a variety of phenomena. When the unnaturalness and the contingency of gendered identities and expectations is exposed, we can then ask: why is that these particular configurations of gender appear in this particular context? Whose interest do they serve? How do they implicate outcomes?'.

Our theoretical framework here is hinged on the critical feminist theory which concerns itself with the 'reformation of concepts and an attack on epistemological assumptions that affect the conception of the position of women in society' (Ityavyar & Obiajunwa, 1992, p. 20). It is a theory of change that holds that 'the long established myths and even jokes that misrepresent or down grade women should be abolished to allow room for a new and a better socialization process in society which will not deny women their rightful positions in society' (Ityavyar & Obiajunwa, 1992, p. 20).

Simply put, conflict can be defined as a disagreement or clash between ideas, principles or people in any situation condition and or circumstances. Interestingly, there is nothing inherently negative or positive about conflict. It is our perception and or views that characterise our interpretation of any conflict. Thus it is an irreducible truth that conflict is as old as man. This is partly because human relationship is necessarily conflictual. The historicity of conflict is such that every historical epoch is replete with conflict at a phenomenal level, a sin qua non to human existence (Isard, 1992, p. 1). This is corroborated by the assertion that conflict is an inevitability arising from choices and decisions in human interdependence (Zartman, 1991, p. 370).

Our conceptual definition of conflict is hinged on the Relational theory of conflict which asserts that violence between groups can be explained when the sociological, political, economic and historical relationships between such groups are taken into account.

Latching onto these theories can help in arriving at a conceptual framework. In a conceptual sense therefore, the paper proposes that all literary texts are a mediation of the dominant history and gender practices prevalent in the society that engendered their emergence. The result of this relationship is an ideology more often than not nuanced in elitist politics which attempts to reify the masses into the world view of the elite. Unfortunately, fewer women are found among Nigerian elites and more often than not it is the male gender that tells their story.

Synoptic Presentation of Women of Owu

The play which is an adaptation of Euripides' play- *The Trojan Women* (415 BCE)-dramatizes the story of the plight of women in war situations. Following an invasion by 'the combined forces of the armies of Ijebu and Ife, two Yoruba kingdoms in the south of what is now known as Nigeria along with mercenaries recruited from Oyo refugees fleeing downwards from the Nigerian savannah land, sacked the city of Owu after a seven year siege' (*Women of Owu* vii). The Allied forces invaded the city under the pretext of liberating the flourishing market of Apomu from Owu's control. Owu closed its gates. In the third year worsening drought ensued bringing hunger and starvation. In the seventh year Owu inhabitants sensing that the Allied forces had departed open the gates. The Allied forces lurking in the bushes gain entry into the city killing all the males - (adult and children) and carried the females into slavery.

Ideology and the Story of African Woman in Conflict Situation

From a historical point of view, Eagleton (2012) has attempt to roughly periodize the chronicle of modern literary theory in three stages, progressing from,

a preoccupation with the author..., an exclusive concern with the text..., and a marked shift of attention to the reader over recent years. The reader has always been the most underprivileged of this trio-strangely, since without him or her there would be no literary text at all (p. 74).

The interesting aspect of storytelling is the simple fact that every story is by somebody and for somebody. More often than not, the strategic position of the reader is neglected in favour of the story teller and the story itself. However, just as Heraclitus, the Greek philosopher once stated, that no one can step into the same river twice, I dare say that no two stories even of the same event are ever the same. What this presents is an ever changing sameness/scenario.

This dialectics of ever changing sameness is partly a function of ideology and or aesthetics. In fact, every story (retell) is an aesthetic event not only because of what is selected or deselected, but moreso because of how the retell is presented. The result of course is a mediated equation. This is partly so because,

the way in which the ideology of a class or other group is expressed in literature... will be affected or mediated by two sets of condition at the aesthetic level... (i) the conditions of production of works of art and (ii) the existing aesthetic convention (Wolff, 1993, p. 61).

While the second condition might not interest us much here, the first does a great deal. The fact that literary productions today in Africa (and stretching back to about seven decades ago) have been caught in the web of capitalist commodity production is indicative of the political system on which the economy is predicted. Money therefore is the basis of every transaction. The result is that one might have a good story to tell but without money that story might never see the light of day in terms of publishing. Just as Wolff opines: "What gets produced and performed and received by audiences is often determined by straightforward economic facts" (1993, p. 46).

Every storytelling is therefore a retell, often coloured (at other times marred) and mediated by ideological, sociological, historical, demographic and even economic factors. The thing is, the one who narrates the story almost always tells it with a tint of subjectivity. S/he can decide to edit one aspect out rather than the other for whimsical reasons. S/he can decide to tell the story with emphasis on this rather than that aspect. Well, that is as far as telling the story goes. The other issue is getting it published and the wherewithal to do this. And above all, how does the storyteller tell the story? How does s/he depict in the story the inner (ill) logic of society? For Aristotle then, it matters a great deal how the poet/dramatist depict the workings of society. And this, according to Aristotle, can be in one of three ways: '...we must represent men either as better than in real life, or as worse, or as they are' (Koss, 1997, p. 3). This implies that the artist '... must imitate either things as they are, things as they are thought to be, or things as they ought to be'(Koss, 1997, p. 3).

Incidentally, in Africa, both male and female authors have attempted to retell the story of the African woman. Nevertheless, the pendulum swings more in the direction of the male than the female. For example, the pioneer generation of Nigerian dramatists- Wole Soyinka, Ola Rotimi, John pepper Clark-Bekederemo and Zulu Sofola- have presented women in not too good light. Soyinka, in *The Trial of Brother Jero*, portrays Amope, Chume's wife as a troublesome and nagging woman. Clark-

Bekerderemo in *Ozidi* presents Oreame as a witch and diabolical. Rotimi in *Our Husband Has Gone Mad Again* (1977) depicts Sikira and Mama Rashida as traditional and docile women who were ready to tag along the men; Mama Rashida herself was an inherited wife. Flora Nwapa (2010) corroborates this when she asserts that;

Nigerian male writers, such as Chinua Achebe, Cyprain Ekwensi, Wole Soyinka, J.P Clark and Elechi Amadi, have all in their earlier works played down the powerful role of women. Unlike Peter Abraham, Ousmane Sembene, Nigerian male writers have in many instances portrayed women negatively or in their subordination to men. Ekwensi's Jagua Nana is a prostitute. Wole Soyinka's Amope is a ceaselessly nagging woman who makes life intolerable for her husband. ...J.P Clarks's Ebeire entices her husband's younger brother into a sexual relationship. The focus has always been on the physical, prurient, negative nature of woman (p. 361-362).

The second generation of Nigeria's dramatists comprises the likes of Femi Osofisan, Bode Sowande, Segun Oyekunle, Kunle Omotoso, Tunde Fatunde, Tess Onwueme, etc. Here Osofisan and Onwueme stand tall. However, Osofisan remains the most prolific. Most of his plays have been staged widely across the length and breadth of Nigeria and Africa. Interestingly, Osofisan has also been outstanding in the manner in which he continues to retell the story of the African woman by subjugating historical materials to contemporary realities. These dramatists were affected by landmark events such as the Nigerian civil war which has led critics to label them as the post war generation. But more importantly, these writers have a philosophical outlook that was shaped by Marxism. No wonder their portrayal of women in their works differ somewhat from the first generation. As Nwapa (2010) asserts; 'Recent changes in Nigeria- the 1967-70 civil war, economic changes and an emphasis on the education of women- have affected men's views about women. Women have started to redefine themselves; they have started to project themselves as they feel they should be presented' (p. 359).

Osofisan's Marxist ideological leaning often leads him to reconstruct and reinterpret history in favour of the oppressed class which often includes women. For instance, in *Morountodun* (1982), Osofisan;

...reinterprets the Moremi myth and legend of the past...through the character of Titubi who plays the role of Moremi. Titubi's heroic impulse is spurred by the courage of Moremi who, according to the legend, had earlier ensured victory for Ile-Ife people through her espionage activity. Titubi, unlike her legendary role model Moremi, fights for the government forces with aim of suppressing the peasant farmers' revolt against injustice and oppressive system of taxation (Awodiya, 1995, p. 40).

However, through careful characterization, Osofisan makes Titubi experience the pains and sufferings of the peasant farmers which led her to commit class suicide-'unlike the mythical Moremi who fought for the government forces to ensure victory for her ruling class' (Awodiya, 1995, p. 40). Titubi switched camp from her rich and luxurious upper class to the proletariat class of the suffering and oppressed peasant farmers.

It is significant to note Awodiya's arguments that: 'There are three groups of women in the plays of Osofisan' (1995, p. 89). To this end, it is important to consider these 'groups' even if it is to see the application to the text under review. According to Awodiya,

The first group of women comprises young and progressive, positive and revolutionary women. Examples of these radical women are Funlola and Yajin in The Chattering and the Song...the second group are the socially conscious liberal women who are eager for social reform. They do not lead the revolution themselves but are being led by more radical fighters for social justice. ...Examples of the liberal women are the market women in Once Upon Four Robbers, the peasant women in Morountodun, ... The third group include conservative, reactionary and usually old women who lack consciousness and are adamant to social change. Such women who have tendencies to oppose or resist social transformation are Alhaja Kabirat in Morountodun, Tolu in Who is Afraid of Solarin?... (1995, p. 89).

It can be argued that over all, what Osofisan has been able to achieve with these groups of women is (at various times), portraying them as they are, as he thought them to be, and as they ought to be. This is a fair attempt to mediate the variegated narratives of women into our national consciousness. In addition, Osofisan has also shown that women are capable of initiating, engaging and sustaining the quest for social justice and transformation. Thus, women in Nigerian societies are strategic stakeholders to be reckoned with since they can be heard and seen as well.

Awe (2001) has succinctly chronicled Nigerian women and this is worthy of mention. Although Awe's book is not a poem, however, in a sense, it equals Homer's *Odyssey* (700 BC) and *The Iliad* (750 BC), especially in the wealth of the history of Nigerian women, it attempted to document for posterity. This book is also worthy of note because it can serve as a source material for dramatists to continually draw upon and present a more accurate picture of the positions of these women in the history of Nigeria and their contributions to national development as a whole.

The various themes then in Osofisan's *Women of Owu* reflect and mediate the mutating contexts of the African woman's challenges in conflict situations before colonialism, during colonialism and in today's neo-colonial Africa.

Thematic Analysis: *Women of Owu*

There are numerous themes in *Women of Owu*. A rundown will include,

- The after-effects of war (desolateness, rape, displacement, slavery, degradation, death),
- Greed/lust for power,
- Sympathy for women,
- Conceit of the gods /demystifying the powers of the gods,
- Patriotism

From the avalanche of themes, we shall zero in on just three: conceit of the gods/ demystifying the powers that be, the after effects of war and the sympathy for women. Why? For the simple fact that these three themes elicit the strength and thrust of the play not to mention the aesthetic dimension Osofisan brings to the table of his adaptation.

The most glaring theme is the after-effects of war. It is within this matrix that we witnessed rape, displacement, slavery, degradation, and death of the citizens of Owu. Using a late point of attack approach, the play opens and ushers us straight into the thematic context. Owu has been sacked by the Allied Forces. The stage directions hint us thus:

The night before, the King, Oba Akinjobi had fled from the town, with some of his high chiefs and soldiers, leaving his family behind. The Allied Forces slaughtered all the men left in the town including the male children; and only the female children and women have been spared and made captives. ... Visible in the background is the city itself, in ruins, and smoldering (Osofisan, 2006, p. 1).

It is instructive to note that although in the nihilist rampaging of Owu, women and female children were spared, we must not lose sight of the fact that earlier, the King, his Chiefs and some soldiers had fled leaving the women and children begind. In addition, the fact that the women and children were taken into scaptivity is not a pleasant experience altogether. However, the two women from Owu who were accosted by Orisa Anlugbua and the dialogue that ensued is revealing. Anlugbua-"a former war leader, son of Oba Asunkungbade, ancestral founder of Owu Ipole, now also deified as *Orisa*" (Osofisan, 2006, p. ix). The conversation between them at the end only serves to highlight once again Osofisan's disdain and rejection of the superiority of the gods over human beings. At first the women are unaware of the man they are conversing with, to them he was just a mere stranger:

Anlugbua:

Tell me, dear women-
You seem to come from there-
What's the name of the city I see
Smouldering over there?

Woman:

Stranger, you don't know? Look at my tears!
That was once the proud city of Owu, reduced to ruin
yesterday...

Anlugbua:

Yesterday!...

Woman:

Yesterday, old man!
For seven years we had held them off.
These invaders from Ijebu and Ife, together
With mercenaries from Oyo fleeing south from the
Fulani forces. They said our Oba
Was a despot, that they came to free us
from his cruel yoke...
Until yesterday, when a terrible fire engulfed the city
and forced us to open our gates. That was how
they finally gained entry and swooped on us...
And before our very eyes here,
They rounded up all our husbands and
Brothers and sons, and slau-
(Stops, choked by emotion)

Woman:

They slaughtered them! All!...
And shame, oh shame! Our women were seized
And shared out to the blood-spattered troops
To spend the night. Only some of us-
Were spared-those of us from the noble houses
And others whose beauty struck their eye:
We are being reserved, they say, for the Generals (Osofisan, 2006, p. 1-2).

The glaring revelation here is the inability of the gods to come to the rescue of the people (in their time of need), who have worshiped them for ages. It is unimaginable that for seven years the gods did nothing to lift or deliver the people of Owu from the siege. Indeed, it's a shame that even after the slaughter and pillage, many women were raped and children (especially male), slaughtered. Then the women were filled with awe when they realized who Anlugbua really is;

Anlugbua:

My words were clear enough, I thought!
When any grave danger threatens the town,
How was it that no one remembered?

Woman:

You... You... Who are you?

Anlugnua:

Three times, I said! Call my name
...and I shall be back,
Sword in hand to defend you!

Woman:

Sword! That would have served little purpose
This time, I tell you! Because- eh! Yeh!... Yeh!
What did you say? You... you... mo gbe!...
Is it you... ? Have I stumbled upon...
Impossible! My eyes have not seen a...
No! Impossible!

Anlugbua:

Calm yourself, my dear women.
You have nothing to fear!

Woman:

It's not you, is it? Let me not
Go blind today! It's not you, Orisa Anlugbua?
Not our ancestors they so talked so much about?

Anlugbua:

I said-

Woman:

But-(Screams) So it's true, ehn?
Help! Help! The god is here!...

Anlugbua:

(Hard)

Sh! Quite, I said! Calm yourself!
I don't want to be announced! Not yet!...

Woman:

(Cowed)
Yes…yes…

Anlugbua:

So go on, tell me all that happened here.

Woman:

(Whimpering with fright)
Ah…ah…(Osofisan, 2006, p. 3-4).

All the awe however soon turned to scorn when it dawned on them that the god that was so much talked about has failed them:

Woman:

Maye besieged our city for seven full years,
Because of woman, and would not go away!
For seven full years, the people of Owu
Suffered and refused to open the city gates.

Woman:

Seven years without rain they were, seven years,
Of failed harvest. All of those terrible years
Where were you, Anlugbua?

Anlugbua:

You did not send for me! You know
The oath I made forbade me to return
Here, unless you sent for me!

Woman:

After three years, the city
Began to starve of food and fruit,
And the streets stank of disease and death
But we never seized
To worship you, or pour libation at your shrine!
Where were you, Anlugbua?...

Anlugbua:

Well, it's all over now
The Allies have got what they wanted
I've come too late (Osofisan, 2006, p. 7-8).

Eventually, Anlugbua's subterfuges could not hold water; even the gods had to concede defeat no matter who is to be blamed. But that is not the worst yet. What follows is a denigration of the revered position of the gods; one of Osofisan's hallmarks:

Anlugbua:

It's very sad, my dear women! But
Still with all your tears, I, Anlugbua,
I am the real loser here,
Gods do not cry. But that only makes the pain deeper still.

Women:

So what do you want? What have you
Come back for? So that we may pity you,
We ourselves who ache for consolation?

Anlugbua:

I ask you-without a shrine, without worshipers,
What is a god? Who now will venerate us?
Who sing our praises among those ruins?

Woman:

Go back to your heavens, Anlugbua
And learn also how to cope with pain.
If only you gods would show a little more concern
For your worshippers!...

Anlugba:

I confess, I am broken. Farewell my lost city.
Farewell my dear women, whose arguments
Shame me.... (Osofisan,2006, p. 8-9).

Here, Ososfian once again demonstrates his dexterity in interrogating Yoruba myths/legends and traditions in an attempt to re-assert the strategic position of human beings in defining their reality and or destiny.

The second theme in focus is the after-effects of war. It is bad enough that the war took place, however, what followed after was even more devastating. The play is replete with the after-effects of war majorly in terms of desolateness, rape, displacement, slavery, degradation, and even death. In all these, it is the women who are left to cope with them.

Woman:

Liars! You came, you said,
To help free our people from a wicked king. Now,
After your liberation, here we are
With our spirits broken and our faces swollen
Waiting to be turned into whores and housemaids
In your towns. I too I curse you!

Erelu:

Savages! You claim to be more civilized than us
But did you have to carry out all the killings and carnage
To show that you are stronger than us? Did you
Have to plunge all these women here into mourning
Just to seize control over our famous Apoma market
Known all over for its uncommon merchandise? (Osofisan,2006, p. 12).

These atrocities demonstrate the contradictions of war and only go to show that behind any war lie deep seated personal ulterior motives.

The third and last theme we shall consider is that of sympathy for women. The courage of the remaining women of Owu is commendable. No matter how they wanted to disagree with the gods, in the end their only recourse was to the gods, who once again refused to help them in the final ritual of sending the spirit of the dead to the here-after. However, the recourse to the gods was a bold step even if it is to fulfill all righteousness. So if this courageous posture painted the women in good light – a humanistic view that counters the hegemonic stance of other dramatists.

The war may have been over and all the damages done, yet the rituals of life and death must continue and our spirits must move on. The implication of the courage displayed by the women of Owu is that this kind of never-die spirit should resonate in today's African woman.

CONCLUSION

We have considered Femi Osofisan's playtext *Women of Owu* adapted from Euripides' Greek version *The Trojan War* (315 BCE). The arguments we have presented only goes to show that in any story, the storyteller has the advantage to tell the story in one of three ways ala Aristotle's proposal. Interestingly, the manner in which Osofisan has chosen to re-tell this story is a kudos to the African woman in terms of the amount of sympathy they attract and the humanist perspective which Osofisan brings to bear on the work. In the end, the message cannot be mistaken: the remnants of the women of Owu (even as they go into slavery) will start a new Owu wherever they dwell and a new Owu will take root albeit in several places. So all hope is not lost!

REFERENCES

Awe, B. (2001). *Nigerian women: A historical perspective*. Ibadan: Bookcraft.

Awodiya, M. (1995). *The drama of Femi Osofisan: A critical perspective*. Ibadan: Kraft books.

Carpenter, C. (2002, Autumn). Gender Theory in World Politics: Contributions of a Non-Feminist Standpoint. *International Studies Review*, *4*(3), 153–165. doi:10.1111/1521-9488.00269

Clark, J. P. (1962). *Ozidi*. London: Oxford University Press.

Eagleton, T. (2012). *Literary theory: An introduction*. Minneapolis University Press.

Emenyonu, E. (Ed.). (2000). *Goatskin, bags and wisdom: New critical perspectives on African literature*. Trenton African World Press.

Freire, P. (1970). *Pedagogy of the oppressed*. New York: Continuum.

Isard, W. (1992). *Understanding conflicts and the science of peace*. Blackwell.

Ityavyar, D., & Obiajunwa, S. (1992). *The state and women in Nigeria*. Jos University Press.

Koss, R. (Ed.). (1997). *Aristotle's poetics*. New York: Dover Pub.

Nwapa, F. (2010). Women and creative writing in Africa. In *Landmarks and new essays in literary and linguistic theories and analysis*. Princeton.

Osofisan, F. (1982). *Morountodun and other plays*. Ibadan: Longman.

Osofisan, F. (2006). *Women of Owu*. Ibadan: University Press.

Rotimi, O. (1972). *Kurunmi*. Ibadan: University Press.

Salami, I. (2001). *Emotan*. Jos: Mazlink.

Salami, I. (2004). *Idia, the warrior Queen*. Jos: Sanitez Pub.

Trigiani, K. (n.d.). *Masculinity-Femininity: Society's Difference Dividend.* Retrieved from http://web2.airmail.net/ktrig246/out_of_cave/mf.html

WHO. (2018). *Gender.* Retrieved January 9, 2015 from www.who.int/gender

Wolff, J. (1993). *The social production of art.* New York: Macmillan. doi:10.1007/978-1-349-23041-9

Zartman, W. I. (1991). *Conflict resolution in Africa.* Washington, DC: The Brookings Institute.

Chapter 7
Trado-Modern Religious Ideology Conflict in Nigerian Films:
An Examination of *Scorpion God*

Jovi Chris Okpodu
National Film Institute, Nigeria

ABSTRACT

The concept of traditional religion in the twenty-first century is that which seeks regurgitation. This is so because traditional religion and its values and beliefs are constantly being challenged in the face of Western religion and its values. This chapter is on the contemporary Nigerian movie Scorpion God as directed by Nonso Okonkwo, and its interpretation and presentation of concepts relating to Nigeria's traditional religion. This study takes a look at traditional religion, the patterns, symbols, and values that have remained relevant till date. It examines the theories and ideologies influencing these ways of life in order to correct the errors presented in this Nigerian film. The study reveals that traditional religion as reflected by Nigerian films is a misinterpretation and misrepresentation, and the findings suggest an urgent need for education in order to correct this error in the evaluation to traditional religion and its place in the socio-cultural life of Nigeria.

DOI: 10.4018/978-1-5225-5987-0.ch007

INTRODUCTION

Africa and indeed Nigeria is a highly traditional society and most of her traditions are deeply rooted in religion. Religion permeates the entire African landscape, it is embodied in the way its people live, what they believe and what they do. Despite the diversity and plurality of cultures, the average Nigerian, finds his or her essence of being in the various gods and spirits that are deified in his or her tradition. A cursory look at the various traditional practices would reveal a seeming synergy of beliefs, that there exists a supreme being from whom all living creations and creatures source life.

The African, and particularly the Nigerian, prior to the contact with the Whiteman, has devised various processes of worship which are often mediated by priests and oracles. However, the advent of colonialism redefined not only social relationships but also traditional practices, including religion. The Whiteman came with his 'god' and superimposed same on wherever he found foothold. In Nigeria particularly, the Southern protectorate was the worst hit, as Christianity 'stamped' its people. As a result of massive influx of materials such as Bibles, books, and even the subsequent establishment of Mission Schools across the length and breadth of Southern Nigeria, traditional religion not only became seemingly anachronistic but was also branded as fetish and evil, and the worshippers regarded as 'pagans'.

Elsewhere in Northern Nigeria, Islam, one such foreign religion, gained ground right from the early 16th and 17th Centuries. The Othman Dan-Fodio *Jihad* swept more than a third of the entire region spreading the virtues of Islam. As of today, the Islamic religion has been adopted as not just a mode of worship but a way of life by its adherent. It is somewhat difficult to differentiate between Islam as a religion and the tradition or culture of the Hausa or Fulani group in Nigeria.

Tradition is an established pattern of thought, actions and behaviors of a people. This includes their traditional religious practices and modes of worship. Tradition as a concept is very difficult to define. It permeates every facet of the life of man, as it shapes his values and norms. Traditional religion here refers directly to man's relationship to divine and superhuman powers, bearing in mind all the several indigenous religious forms and systems. Mbiti (1963), says that, "people respond to their spiritual world. This response generally takes on the form of worship which is eternalized in different acts and sayings" (p. 65).

The belief in traditional religion is that man's existence is governed internally and externally by sublime forces belonging to the superhuman or extra-terrestrial realm. The form of veneration he says, may be "formal or informal, regular or extempore, communal or individual, ritual, ceremonial or unceremonial, (Mbiti, 1963, p. 58), through words or deed". It is not in doubt that the subject of traditional religion is quite topical in the 21st Century being a period of great strides in information technology. For example, religions have emerged from countries like China, India, Japan and Europe. The *Hindu* religion and the *Baha'I* faith are fast gaining ground in Nigeria and other African countries. Through the means of information technology including films, these countries export their religion to the rest of the world.

The power of films in influencing society cannot be underestimated. Over the last decade; 'Nollywood' which is the Nigerian version of 'Hollywood' has been branded as the third largest film industry in world. In the last five years, the products of this industry have found their way into Europe and America. This further makes the content of these films a matter of not just great economic concern but cultural and national identity. Collectively, film scholars have agreed that every artist knows that words or films are loaded pistols. If one speaks, one fires. One may be silent but since he or she has chosen to fire, he or she must do it like by aiming at targets, and not at random for the pleasure of the shot going off. (Shehu, 1997, p.176)

This captures the criticism that has trailed the Nigerian Home Video Industry over the years. The lack of a clear-cut ideological predisposition of the industry has made its contributions rather blunt in the universal competition for dominance.

The role of Hollywood in this trend is quite latent. Every American movie carries with it, the American ideology of excellence. Nigerian video films ought to do the same by exporting not just Nigerian culture but Nigerian traditional religion as well. Drawing from the Soviet scene, as explored by Nikita Kruscher, the Soviet cinema is one of the most important artistic vehicles for educating the people in the spirit of communism. There is nothing to compare with the cinema in its power on human minds. The cinema is accessible to all walks of society, to all ages, permeating into the remotest of districts and villages. (as cited in Shehu, 1997, p. 176).

The latent power of film as a tool of re-awakening consciousness and influencing the society makes it imperative to revisit the content of Nigerian films and probe its role in promoting or destroying the ideals of traditional religion in Nigeria.

Dramatic development, including film stories and performance, begins in religion and ends in aesthetics. It is rather worrisome that the basic elemental force that gave birth to drama and later film has been badly represented by its offspring. African traditional religion has been kept at the opposite side of the religious divide. A careful study of most Nigerian Home Video, reveals that evil is associated with these religions, and the so called modern religions continue to gain popularity the world over. Nigerian movie makers have not helped matters either. It does appear that there is a growing misconception about traditional religion if what we see in our own home video is anything to go by. This is the motivation for this paper. The aim is to investigate effectively the content of one Nigerian Home Video in order to show these misrepresentations.

This study is based on two hypothetical assumptions and statements, that;

1. Nollywood has rather prompted a belief in traditional religion that is old archaic and antiquated and that is unacceptable amongst Nigerians as it portrays them as ritualistic and barbaric.
2. In the face of modernity and present social realities in Nigeria, massive religious beliefs, attitudes and practices have been transferred from the traditional to the current, which is Christianity and Islam; even though there are still some adherents in traditional religion and its practices.

Our study in this subject matter will be based on the film *Scorpion god*, as directed by Nonso Okonkwo (2002), while also engaging in library research to incorporate archival resources to the subject matter.

Generally, traditional religion refers to the indigenous pattern of man's relationship to the divine or super-human powers before and after the advent of Christianity, Islam or any other foreign mode of worship. While we accept this definition, it is imperative that for the purpose of this study, we shall see traditional religion as a body or pattern of beliefs, ceremonies, rituals and festivals of a people or community. Mbiti (1963) gives credence to this by saying that "traditional religion permeates all the departments of life. A person cannot detach himself from the religion of his group, for to do so is to be severed from his roots, his foundation, his context of security, his kinships and the entire group of those who make him aware of his own existence" (p.58). Hence, religion not only provides a means of Communicating with extra-terrestrial forces but is central to how Africans or Nigerians identify themselves.

The term "modernization" or "modernity" is then difficult to define. It became a subject of intellectual discourse at the beginning of the 20th Century when humans came face to face with the reality of a dynamic and inventive society. Modernization means different things to different people. To the political scientists, it is a theory which seeks to explore the conditions for economic and political development from a "traditional" to a "modern" society (Encarta, 2008). To the theologians, it is an attempt by a group of scholars and church officials to re-interpret Christianity in terms of the scientific thought of the century. Scottish historian and essayist, Thomas Carlyle (2008) postulates as to the three great elements of modernity and modern civilization as including, the gun-powder, printing and the protestant religion (Encarta, 2008). A careful look at these three concepts point to one thing and that is, that modernization or modernity is synonymous with change within the society. Man lives in a highly dynamic and fluid society and hence, modernization refers to those changes that take place within society which compels man not to remain in his beliefs and attitudes.

SYNOPSIS OF THE FILM 'SCORPION GOD'

The movie centers on two young lovers, *Ude and Iheoma,* who vowed to stay by each by each other matter the odds. Unknowing to *Iheoma*, Ude had compromised himself with another young girl called Asa whom he forcely had sex with before he met her, an act which is sacrilegious and must be visited with grave consequences. The only remedy was to go before the shrine of *Amadioha* (the god of thunder) and confess, thereafter perform a cleansing sacrifice. But *Ude,* in order to keep his pride and maintain his good family name, decided to take a false oath before the priest. Anyone who toes this path risks madness and even death within six market days, as no one lies before *Amadioha* without consequences.

At the end of the sixth market day, *Ude* woke up sound and healthy but his jubilation was short lived, as he was struck by a strange spirit of madness while on a victory tour of the village. While in this state of madness, everyone in the village abandoned him except *Iheoma* who defied even her parents to stay by *Ude's* side. Tragically, *Ude* died a few days after.

Thematic Analysis and Ideological Conflict in the Film 'Scorpion God'

We see what is quite typical of Nollywood movies, an introduction of a new plot, a new story. In a bid to take the story further, *Ikenga* another main character was to begin his journey to the unknown. The priest of *Amadioha* visits him in his house and informs him that his destiny lies in the jungle. What he was to look for there we do not know and we were not told. The movie ends as *Ikenga* is shown a sign which he sought for from the gods – a blazing roar from the depth of the night sky.

The movie opens interestingly by saying that man's destiny often lies on the path laid by 'forces' beyond his comprehension. These 'forces' are often called 'fate'. Sometimes they come after man with venom of an angry scorpion when laws guiding them are violated. When this occurs, we can say that 'fate' has played a vicious game on man.

As such, we begin to see contradictions which typically reveal that Nollywood is yet to grasp the elemental attributes of Africa and African Religion. The understanding we get from this is that a man's fate is consequent upon his actions. The film begins with the saying that a man's destiny often lies on the path laid by 'forces' beyond his comprehension often called 'fate'. Forces and fate first and foremost, are not the same. When something is beyond your comprehension, it means you do not control it, and how it operates within the sphere of human experiences. It is what no man can explain. . In African traditional philosophy, all agree that man's thinking, and actions are governed by an invisible and incomprehensible 'force'. The key concept to

African philosophy is the essence of being. Therefore, force equals being and being equals force. To liken this 'force' to fate as the producers of *Scorpion God* (2002) would want us to believe, negates certain principles of force and may amount to misrepresentation of the facts. Perhaps this further buttresses the need for intensive research by Nollywood movie makers and writers.

On the other hand, destiny or fate is a predetermined course of events considered to be beyond human control. The presupposition that destiny is the eternal destiny of a person is predetermined by the god's unchangeable and unwavering decree. It thus becomes clear that something is fundamentally wrong with the interpretation of the maker of this movie on the issue of 'force' and fate.

One other noticeable case of contradiction in the opening quotation by the *Scorpion God* is the belief that man's actions determine his fate, that fate, especially that which is cruel, comes upon man when certain laws are violated. A man is responsible for his actions and whether for good or for bad, has its consequences. Actions and consequences are part of social norms in all African societies. A man stands up to his actions and whatever happens to him in future is a result of the choices and decisions he makes today. These consequences should not be confused with destiny, just as we cannot confuse 'force' with fate. Destiny is what no man has the power to change or affect.

THE IMPORTANCE OF SACRIFICES IN TRADITIONAL RELIGION

One of the thematic concerns of the *Scorpion God* is the place of sacrifices in Nigeria traditional religion. According to Mbiti (1963), sacrifice and offerings constitute one of the commonest acts of worship among African people. Sacrifices refer to cases where animal life is destroyed in order to present the animal, in part or in whole, to God, supernatural beings, spirits or the living-dead. He advanced four main theories to explain the function and meaning of sacrifices which are; the gift theory, the propitiation theory, the communion theory and the thank-offering theory. Two of these theories were used in the movie which are the Gift and the propitiation theory. *Ude's* father brought live-goat before the priest of *Amadioha* for two reasons; to thank *Amadioha* for giving him a son and to solicit for more children to be born into his family. The other theory was that of propitiation. The gods must be appeased when they are wronged and failure to do this often leads to some disequilibrium and cosmic chaos. The consequences of not appeasing the gods could be either communal, whereby an entire community is judged or individual, where the perpetrator pays for his actions or inactions. In this movie, to put it very mildly, *Ude* consciously wronged the god. He dared the gods and consequently incurred the wrath of *Amadioha*. But a mis-representation had to occur when *Ude* had to lie. All he needed to do was to admit his crime and thereafter perform the cleansing sacrifice. But just like an Ibo proverb says: "when a dog starts playing with death, it hardly perceives the smell of excreta". One wrong led to another, *Ude* lied before the priest thereby inflicting upon himself the wrath of *Amadioha*. A man must stand up for his actions and be ready to accept the consequences. *Ude* thought he could outwit the gods but maybe, he never knew that "what an old man sees sitting down, a child cannot see even if he climbs the top of an Iroko tree". What one see's here is the convenience of mis-interpretation for the elongation of cinematic time.

The Inalterability of Destiny

One of the significations that we get from this movie is that a man's destiny is not subject to change. Though this may sound quite absurd within the circles of the modern period and the social realities of our time, it is a philosophy that goes with traditional religion. In the *Scorpion god, Ude* went to the shrine to challenge the gods to revert his destiny when it became clear to him that disaster loomed around the corner following his false oath. He was confronted by the priest and the following dialogue ensued:

The Priest
Young man! What are you doing here?
Ude
Old one! I came to change my destiny…
I have the will to change my destiny.
The Priest
Your destiny has been sealed.
Ude
Lies old one.
The Priest
You lied before Amadioha.
Ude
Nothing can stand before a man who has strong will.
The Priest
Your fate has been sealed.
Ude
Nobody has the right to decide my fate and destiny.
(Scorpion God, 2002)

With this, he proceeds to the shrine to tear off the barricades but was engulfed in a seeming electric charge that threw him off the shrine to a nearby bush. He ran away and never went back.

Our present social realities have proven that man cannot by himself change his destiny. Western religion such as Christianity subscribes to this ideology even though some modern day gospel preachers have cashed in on the ignorance of their adherents to emphasize that humanity possesses the power to alter its future. Western ideologies of the modern period believe that by a conscious effort, man can make better his lots hence nobody is born to be poor or rich. One only finds himself or herself on either side of the pole by the decisions and actions we make. These decisions and actions are still subject to some divine source. Therefore, man cannot arrogate to himself godly powers and does need every divine intervention in his affairs.

Perhaps this is where traditional African or Nigerian religion needs to be revisited objectively. What the *Scorpion god* typifies in this regard raises a lot of issues of intellectual discourse. Destiny by every standard in traditional modes, beliefs and practices is not what human beings have all the power to decide and change. A man does not have the power to decide when he should die just as he never was part of decisions reached on when he should be born. This alone reveals that man does not have all the answers to life's unending questions.

THE REVERENCE OF MARRIAGE AND OPPOSITE SEX RELATIONSHIPS

Marriage rights from ancient times have remained rather complicated as it is laced with economic, social and religious implications and perspectives. In Africa and in Nigeria in particular, it remains the pivot and significant unit of society. Marriage is highly revered and in most instances sacrosanct to the people. Therefore, all religions have high regard for the marriage institution. Every work of art especially the Nigerian Home Video has had their works revolving around marriage and the relationships between the opposite sexes.

Marriage in Africa remains a mystery with lots of religious undertone. It captures the entire cosmology of the African people and brings to life the people's deep spiritual essences and experiences. It is a mystic union, a transition which encourages amicable co-existence of all forces. Soyinka (1976), calls this the dark center, "where mysterious continuum of transition and inter-transmutation of life, both essential and materials occur".

Therefore, marriage presents to us continuity in movement, a perpetual transition and cyclical in nature, eternal, yet not dichotomous. It is a meeting point between the past and the future, where the past is transmitted into the present, and the present is in passage to the future.

The institution of marriage was highly revered as we see in the movie, *Scorpion God*. Sex remains an exclusive preserve of marriage and only those who are rightfully married must indulge in the act of sex. Young boys and young girls were and are forbidden from premarital sexual experience, and anyone who indulges in it risks the wrath of *Amadioha*. This was what caught up with *Ude* as he was said to have had an affair with *Asa*. Suffice it to say that in those days, promiscuity was greatly minimal even though young girls and boys have very little clothing which barely covered crime their breasts and waists. They had great respect for the gods, and marriage was paramount in their minds. *Asa,* one of the characters in the movie had no problems with *Ude's* lies before the shrine, because he felt that if the young man admits his, no man would ever set eyes on her again, not to think of marrying her. Such was the reverence they gave to the institution of marriage as ordained by the gods.

CONCLUSION

This study has proven that traditional religion is neither evil nor is it fetish, and that there is every reason for a re-writing and re-interpretation of those films as presented by Nollywood, for its true and proper meaning, not only for the education of the modern Nigerian, but for the education of the imperialist west as well. Film is an information item as it is an export product, and it speaks a lot about where that film is coming from, and since information drives the world in the 21st century, Nigeria in particular and Africa in general cannot afford to sell the wrong information about itself. Finally, and most importantly, members of the academia need to intensify research effort in the area of the traditional ideals and values of Nigerian societies.

REFERENCES

Mbiti, J. S. (1963). *African religions and philosophy*. London: Heinemann Educational Publisher.

Microsoft Encarta, . (2008). [Computer software]. Redmond, WA: Microsoft Corporation.

Okonkwo, N. (Director). (2002). *Scorpion god* [Motion picture]. Nigeria: N.A.

Shehu, B. S. (1997). Film, Culture and Democracy in Nigeria. In Y. Nasidi & I. Igoli (Eds.), *Culture and democracy*. Zaria: ABU Limited Press.

Soyinka, W. (1976). *Myth literature and the African world*. Cambridge University Press.

ADDITIONAL READING

Dipio, D. (2007). *Religion in Nigeria home video. Westminster paper in Communication and Culture* (Vol. 4). London: University of Westminster.

Paden, E. W. (2008) “Religion” Microsoft Student 2008 [DVD] Redmond, W.A: Microsoft Corporation.

Traore, B. (1972). *The black African theatre and its social functions*. Ibadan: University Press.

Chapter 8
Vigilantism and Banditry in Southern Katsina Emirate, Northern Nigeria 1987-2015

Imbrahim Sani Kankara
Umaru Musa Yar'adua University, Nigeria

ABSTRACT

The focus of this chapter is armed banditry by the Fulani pastoralist and vigilante activities of farmers. The banditry in Pauwa district is shaped by the nature of the environment and the occupational activities of people in the area. The area is approximately 64.78 square miles. The recent rise in the activities of farmer vigilante groups in Southern Katsina emirate was associated to high activities of cattle rustlers and banditry by the Fulani pastoralist. These activities were identified to be the causes of armed violent conflict associated with farmer and herdsmen conflict. Thus, to underscore the nature of conflicts between the pastoralist and farmers in the area, the study has adopted structuration approach in describing the role of structure and individual agents in the conflict.

DOI: 10.4018/978-1-5225-5987-0.ch008

INTRODUCTION

Historically some people live by agriculture; the cultivation of vegetables and grains: others by animal husbandry, the use of sheeps cattle, goats, bees, and silkworms, for breeding and for their products (Khaldun,1967). The cultural relations between the Fulani pastoralist and other communities through the Nigerian federation represent a focal point and important component of relationship between the teaming communities that are either farmers or postoralist. The sociocultural pattern of relationship between the Fulani pastoralist and Hausa farmers can be described as one of the best relationship that exist between the different Nigerian ethnic groups. This relationship in the recent past deteriorated due to incessant attacks on villages, cattle rustling, kidnapping, maiming and extra-judicial and violent response through vigilante activities. This is all associated with the banditry nature of activities of both the farmers and the pastoralist. Banditry is a specific phenomenon that exists within the socio-economic and political parameters. Banditry featured prominently in societies where the law takes the form of blood feud. It is a strong security challenge (Hobsbawm, 2012). The banditry in Pauwa district is shaped by the nature of the environment and the occupational activities of people in the area. Presently, the district is in Kankara Local Government Area of Katsina State. It is one of the well known districts of Katsina emirate, in the history of iron smiting in the 19th and 20th centuries. The location of the district along the fringes of Dajin Rugu and Yar'santa forest reserve stands as an important critical parameter in this context. It however marked a major distinctive difference between the area and other districts.

The recent rise in the activities of farmer vigilante groups in Southern Katsina emirate was associated to high activities of cattle rustlers and banditry by the Fulani pastoralist. These activities were identified to be the cases of arm violent conflict associated with farmer and herdsmen. The farmers and pastoralist related violence has been on the increase in Nigeria. According to Olayoku (2014) about 615 violent deaths have been recorded by the database of Nigerian Watch and there have been about 61,314 fatalities that can be described as violent. According to the scholars, in 2008, there were 31 reported cases of cattle conflict while in 2009 the fatalities soared to 83 all occurring most primarily in the northern states of the country.

Many analysts tried to attribute the recurrent conflict to ethno-religious parameters which define the nature of Nigerian state as a multi ethnic and religious state. The farmer-pastoralist conflicts in Katsina are essentially economic conflict, irrespective of the religious, cultural and political colorations (Abbass, 2012:191). The conflict is related to population and growth and environmental degradations. The farmer-herdsmen conflicts involve Fulani and Hausa famers, Fulani vs Tiv Farmer, Fulani vs Jukun Famers, Fulani vs Birom etc. Although, Fulani herders and farmers have for long coexisted in symbiotic relationships that have persisted through both peaceful and contentious encounters, reports of violent clashes between these two groups are becoming more frequent in the recent Nigerian History (Oyama, 2014:104). This has been the major source contention in Adamawa, Plateau, Nasarawa, Bauchi, Kaduna, Benue, Taraba, Katsina and Zamfara States (Oyama,2014).

With rapid population growth in both urban and rural areas in the country there is increasing demand for food in the country. Similarly, the Farmers and pastoral people are increasingly trying to make use of the same land in order to meet the growing demand. The demand of cities for crops and meat induces many producers to maintain large herds and to expand farmland. The availability of road transportation enables delivery of even fragile resources to remote markets and locations that are difficult to reach. It is along this line that conflicts erupt between the two interest groups. The farmer in his quest to meet the growing population demand expands farm hectares into the grazing reserves. While, the quest for greener pastures by herdsmen usually brings them in contact with the sedentary population who are involved in crop production (See Oyama 2014). The struggle between these two interest groups, coupled with growing breakdown of law and order in the country led to continuous clashes between the two.

Four factors in particular account for this impressive development in the area. The first is the location of the area. The area is approximately 64.78 square miles (Kofar Sauri,1986). The area situated on the fertile loamy soil of southern Katsina state which is one of the most agriculturally productive regions of Katsina emirate. The landscape of the area could be divided into two broad categories, the rich loamy soil in the central and eastern part of the area comprising the village areas of Ketare, Dan Maidaki, Tudu, Gurbi, Jan ruwa, Kuka Sheka, Kankara, and the western part could be described as mountainous area. These include the areas of Yar Tsamiya, Pauwa, Zango, Katoge, Sherere. Also the thick forest of Pauwa situated south of *Dajin Rubu.* The area is dotted with some rocky hills especially in Pauwa, Zango and the areas that constitute the western part of the district. The Yar tsamiya areas also have such hills, there are also occasional mountain around Pauwa, Zango, Bakkai, and Burdugau and Yartsamiya Jino. These rocks and Hills have played a significant role and to some extent serve as places of refuge for some of the cattle rustlers in the area as will be examined subsequently. The second important factor is the population growth of the area. The area have a total population of 243,259 according to 2006 population census (NPC Gazette,2009). This large population represents not only a veritable market potential for industries but also an abundant source of labour both skilled and unskilled. Secondly, no other area in Katsina has natural resources like Pauwa district. The whole area enjoys fairly adequate rainfall conditions varying from 20 to 50 inches. Temperatures also are high throughout the year so that agricultural production is possible almost everywhere within the area. The population in the area supports the existence of many vigilante groups that exist in the area.

A third factor accounting for the conflict between the pastoralist and farmers in Katsina emirate is the relative availability of grazing routes that transverse the area connecting the district with major forest reserves of Dajin Rugu through Yar'santa and Yar'tsamiyar Jino. In the past, the area is known to be accommodating different pastoralist groups that mostly stayed on transit during their annual seasonal migration to southern part of the country during the dry season, or moving back during the rainy season.

The fourth and final factor stimulate the conflict in the area especially during the dry season is the presence of earth dams that were constructed to support the villages means of water supply. These earth dams were identified as major source of conflict between the pastoralist and the farmers. The farmers always accuse the pastoralist of trespassing into their farmlands when going to water their cattle in the dams during rainy season. While in the dry season the farmers accused the pastoralist of over using the water and at the same time contaminating their drinking water. These had been the sources of conflict in many parts of the district in the areas of Bagoma, Nayadi, Unguwar Sarkin Noma, Yar Santa Jeka Areda, Yar'tsamiya, Jandawo etc. Furthermore, the presence of these dams as a factor in the crises represents a social cost of the earth dams. The aim of this paper therefore is to explore the nature of conflict between the Fulani pastoralist and farmers in Pauwa district so as to show how individual agency played a leading role in shaping the conflict over the years

Theoretical Framework

This paper is guided by the Anthony Giddens structuration theory. Giddens published *New Rules of Sociological Method* (1976) and came up with the idea of structuration. Giddens is identified to be among the world's most cited sociologists (Jones & Karsten, 2003:6). The theory provides a structural framework to explain the conditions that govern the continuity and change within social structures, and therefore maps the reproduction and development of social systems (Bratt, 2014:6). The concern of the theory is with 'order' as the transcending of time and space in human social relations (Giddens, 2009:87b). The theory was originally drawn from the ideas of Durkheim, Marx, and Weber. The structure is described to cover three dimensions of signification, domination and ligitimation, while agency is described as human actions covering three elements of reflexive monitoring, rationalisation and motivation of action (Lamsal, 2012:115)

In particular, his structuration theory continues to be drawn on in a wide range of theoretical and empirical studies. Structuration theory may be seen as an attempt to resolve a fundamental division within the social sciences between those who consider social phenomena as determined by the influence of objective, exogenous social structures and others who see them as products of the action of human agents in the light of their subjective interpretation of the world (Jones & Karsten, 2003:7).

Giddens viewed structure and agency not as independent and conflicting elements, but as a mutually interacting duality (Lamsal, 2012:113). Social structure based on his argument are being drawn on by human agents in their actions, while the actions of humans in social contexts determine the production, and reproduction of the social structure (Jones & Karsten, 2003:8). The relation between structure and agency in human society is described as duality of structure being linked by the modality as locus of interaction. The duality of structure represents an ability to link the past action with the present context and what is expected in the future.

Structuration theory's emphasis on praxis begins from the intuitive and appealing insight that whatever types of social events or forms of structured collectivities may arise or change in a given culture or historical era, these types of events or forms of collectivities are generated in the course of social conduct and through the consequences of this conduct (Cohen, 2004:811).

Giddens believes that history offers too many different cultures and civilizations with diverse historical trajectories to postulate the primacy of one set of structural factors over all others. Thus, postulations on *authoritative resources* provided by this theory as part of the methods of structural analysis are inspirational to this particular research. This will inform our understanding of the role of political elites in shaping the activities of *Kauraye* and the circumstances and manner of *Kauraye* activities in the society

Also the theory is relevant to the research because of its prepositions on social norms and the applications of sanctions by law enforcement agencies in order to promote conformity in the society, and how society can be protected against non-conformity (Giddens, 2006:796). The activities of pastoralist and the vigilante as agents in society according to the theory, involved both negative and positive activities which are at the same time sanctioned through either levied formally or informally by structural state agencies. The formal levies, according to the proponents of the theory, are levied by the appointed authorities such as the police, Army and other related law enforcement agencies. Thus, the essential theoretical prepositions that guide this research includes, knowledgeability, historicism, dialectic of control, modernity, modalities, domination, signification and legitimation. The research is further guided by the human utilisation of authoritative and allocative resources in the production and reproduction of social action with the time-space.

An Overview of Farmer/Pastoralist Conflicts 1987-2015

It is important to note that while examining the nature of conflicts between the pastoralists and farmers, it is pertinent to pay attention to the social class in the area. Social class represents a sum total of individuals bound together by certain interests which as a class they try to preserved and protect (Nkurmah, 1980:17). Social class is often associated with habits, institutions, dress and organisations. Basically, Katsina is a stratified society since pre-colonial period, with aristocratic class and the wealthy enjoying high degree of prestige, wealth and power. Giddens argues that "...rapid economic transformations greatly affected the measurement of class..." (Giddens, 2009:308d). The economic and political transformations in Nigeria greatly influence the social class over the years. This is why it is pertinent to note that post-independent class structure is difficult to describe because of the nature of inter-relations between the various classes. In other words the social class in Nigeria is not closed, and social mobility among Nigerian social classes is not forbidden. This therefore becomes a dominant the feature of post-independent social class. The society can be categorized into three different social class component groups of upper, middle and lower. The majority of the cattle and farm land in Katsina are owned by the lower class peasantry and the upper class covering the ruling elites both traditional and modern. This therefore made these social class a stakeholders in the conflict.

The social and historical cooperation between the farmers and pastoralist in many communities in Katsina is surely broken down by the continuous tensions created by the farmer's militia vigilante and the Fulani cattle rustlers. Cattle rustling and banditry have for many years remained the most teething problems bedeviling the people of southern part of Katsina state especially communities surrounding Dajin Pauwa, Yarsanta, Daji Rugu and Mara areas. From personal experience between 2011 and 2013 many people across both farmers and Fulani pastoralist lost thousands of cattle to the bandits. The incessant attack by the bandits led to the emergence of different vigilante group united under the *zaman lafia noma* groups. The zaman Lafia group in Kankara LGA was headed by Adamu Koshe an old ally to late Isan Gona who used to be the National leader and founder of Zaman lafia with his base in Kokami in Bakori LGA. The role of Koshe in the Vigilante activities makes him very famous across the region and beyond. He was known for his charms *Asirai* and *Maganin Bindiga* (traditional immunity against gun shorts). His efforts indeed attracted the attention of the bandit groups and his eventual assassination having identified him as a major threat to their notorious activities.

The rise in the Banditry and cattle rustling is attributed to different dynamic factors such as; lack of grazing reserves due growing demand for faming land, the emergence of new cultural practice associated orchestrated Fulani youth groups with high material life, and Rising meat prices for consumers are thought to be fuelling an explosion in the theft of sheep and cattle with tens of thousands of animals being stolen from farms each year. The emergence of Banditry is on connected to the rising conflict between the Farmers and the pastoralists in the area. It is now a known fact that the major challenges facing agriculture and rural development in Nigeria is the problem associated with farmer-pastoralist conflicts for arable land. Increasing frustration and impoverishment of farmers occasioned by perennial and extensive farm plot destruction and the ensuing bitter conflicts are eroding the gains of agricultural and rural development interventions (Adelakun, Adurogbangba & Akinbile, 2015).

The farmers and Pastoralist livelihoods support millions of Nigerians and form an important part of the rural economy and society. While these two essential components of rural society have enjoyed cooperative relationships in many parts of the country, such coexistence is now strained or has broken down in many communities. The pattern and scale of farmer-pastoralist conflicts varies across Nigeria, and in some places the two groups do still accommodate each other. But in many states clashes have intensified killing and destruction have reached unprecedented levels in major flashpoint centres in both northern, central and southern Nigeria, and tensions have now extended to the south. Theses had covered the states of Oyo, Ekiti, Rivers, Cross Rivers Kogi, Nasarawa, Benue Taraba, Zamfara and Katsina (Higazi & Yousuf, 2015). The clashes between the two groups in Nigeria and other West African countries have a common have a common feature. As pointed out by different research reports on the clashes between the farmers and Pastoralists the tensions can be categorized in particular perceptions of why violence happens and who is responsible affects how responses are designed and implemented. Livestock production by Fulani pastoralist for a very long time is associated with poor grazing land, poor management of grazing routes, farmer encroachment, poor livestock feeding pasture and Seasonal migrations.

There have been repeated and bloody clashes between the herdsmen and host communities for several years. These clashes according to BBC (2016) have claimed thousands of lives within a period of twenty years. It is further stated that in 2014, over 1,200 lives were lost which made these set of people the world's fourth deadliest militant group. In February 2016, there was an attack on farmers in Benue State which according to reports killed some 300 people (BBC, 2016). In April 2016, there was another attack in Southern Enugu State which claimed about 40 lives and destroyed properties worth millions of naira leaving thousands fleeing their homes and rendering children orphaned. On the part of the farmers seeking defence, some angry youths, local hunters and farmers in many communities in the country have however acquired weapons against unprovoked attacks by Fulani herdsmen. These people have also vowed to stop what they have described as mindless killings and destruction of their farmlands by rampaging Fulani herdsmen (Punch, 2016). In recent times, the Fulani herdsmen have reportedly killed hundreds in different states of the federation including Plateau, Adamawa, Benue, Enugu and lot of other states (Punch, 2016) the more disturbing aspect is that no prosecution has been made despite promises by security agencies in country.

The Structural Adjustment Programs and Vigilante Activities

The central argument here is to explore how the social problem associated with economic policies anywhere in the world represent a hydra-headed manifestation of pervasive condition ushered in by the existing economic policy, such as the neo-liberal policies of SAP's. The Structural Adjustment Programs (SAPS) in Nigeria was introduced in 1986 by the military authoritarian rulers as an economic plan to re-structure the economy that was in serious crises. Pundits considered policies of the period as one of the major factors that influenced the growth crisis in rural Katsina. The period of Structural Adjustment in Nigeria witnessed inflation, rise in poverty, the decline of wages and retrenchment of workers in both the public and the private sectors of the economy (Osaghae & Suberu, 2005). In Katsina this period has witnessed the emergence of new forms violent crimes and farmer Fulani conflict. Many of the identified vigilante members interviewed traced their involvement into groups to the period of SAP's in the country.

The Structural Adjustment Programs (SAPS) were a special economic plan designed by the IMF and World Bank, which imposed certain conditions on countries that secured loan from them. These programs amount to detail instructions on what the countries had to do, to bring their economies under control of these monetary organisations (Umoden, 1992). The strategy when adequately implemented was geared towards promoting free markets, including privatisation, deregulation, trade liberalisation, reduction of taxes and cutting back subsidies on public provided services. Between the years 1986 and 1988, when these policies were executed as intended by IMF, the private sector of the Nigerian economy actually did grow as had been hoped, with the export sector performing well, but there was falling of real wages in public sector and among the urban classes, along with the drastic reduction in expenditure on public services. The policy resulted in the decline of manufacturing and agricultural sectors, retrenchment of workers, inflation and shuttling of consumer goods (Abdulkadir, 2005).

On the whole, as indicated above, the SAP'S had adversely affected the living standard in Nigeria. Thus, this left a large proportion of youth unemployment, rising wave on poverty and corruption, and incessant increase in the number of school drop outs. The economic hardship created by the SAP's between 1986-1999 greatly resulted in moral degeneration in many families. Parents were forced to abandon their basic responsibilities at home, training and disciplining of their children to struggling for earning daily bread. Radda (2012) argues that the impoverishment caused by the economic policy led to impoverishment of the common man and the inability of small and medium scale industries to break even or even remain in business (Radda, 2011). The impoverishment caused by the policies led to decline in purchasing power of the people, broken homes due in abilities to maintain families which in turn affect children behaviours. With the rising pressure on agricultural land as a result of population growth, decline of the purchasing power in the urban areas people in the rural areas select violent acts and other activities coping up mechanism during the period.

Similarly, it is important to note that the period had also witnessed the emergence of two broad social class identities. There is the elite comprising the political, aristocracy, privilege and merchant classes. On the other hand the masses class comprises the ordinary people, non elite, the poor, the underprivileged and the professionals. (Osaghae & Suberu, 2005). Ordinary people and the non-elite comprise the urban segment such as civil servants, professionals and the working class. While the underprivileged and peasantry comprises the rural dwellers. Many pundits (Abdulkadir, 2003; Osaghae & Suberu, 2005; Radda, 2011) have argued that within this period of SAP's the Nigerian military authoritarian regime succeeded in reducing the middle class in the social structure of the country. Several theories have been advance to describe the negative effects of SAP's on the Nigerian societies. Among such theories is the rise in crime and deterioration of public service. The economic hardships engendered by the SAP had greatly led to widespread of corruption in the public sectors, and an increase in crime rate throughout the country (Fourchard, 2006:7).

The rise in crime rate was a result of new innovations in forms of crimes as part of survival strategies to cope with economic situation of the period. The wide spread corruption among the police and the deterioration of the police public relation led to lost of public confidence on the police. As argued by Olaniyi, the period witnessed the weakening of the police and the rise of organised crimes that continued transforming dramatically leading to the emergence of plethora of non-state policing groups"

(Olaniyi, 2006:2-6). Thus, the rise in various forms of crimes in the country led to the emergence of vigilante groups in many cities, towns and villages in Katsina.

Vigilante represents security outfit mostly composed of volunteers operating under the mandate of communal consensus to fight crimes (Olaniyi, 2005:21). The activities of *Kauraye* can be traced to the role they played as vigilante groups in many rural parts of the state. Thus, according to in-depth interview;

...As the state of insecurity was getting worse nationwide, different vigilante groups emerged in different parts of the states in order to ensure safety of lives and properties. These groups have formed alternatives to the regular police force in the enforcement of law and order, while keeping a watchful eye on the communities they monitor (IDI,Male, Age 65, Kaura, Kunamawa, Rural, 2013).

The *Kauraye* played role in providing security in their various localities against all forms of criminal activities. The criminal activate in the rural side of Pauwa district during the period are mainly associated petty rustling of livestock, seasonal conflict between the farmers and pastoralist. According to sources this development led to the emergence of new groups in areas such as Tsiga, Kakumi, Mairuwa, Maska, Yaribori, Sabuwar Kasa, Jargaba, Maska, Runka, Ruma, Ruwan Godia, Dan Musa, Tashar Biri and Faskari areas. In-depth Interviews conducted with informants have agreed on the role of *Yanbanga* (Vigilante) in the peasant/pastoralist clashes and the fight against cattle rustlers in their localities. Similarly, according to one of the informants;

Quite a number of vigilante groups such as Zaman Lafia, Rahamawa Manoma and a host of others were formed in these areas (IDI,Male, Age 70, Kaura, Kankara, Urban, 2013).

The formation of these vigilante groups can be described as a step towards formalising their activities and gaining relevance in the society. Thus, vigilante groups were formed and organised with some semi formal structures, across the entire state, with each LGA has its own leaders and coordinators at the various wards.

Even though they are not formally recognised by the security agency as vigilante, their activities became relevant to the state structures, because of the role they were playing in mitigating communal clashes, cattle rustling and fight against undesirable elements particularly in the rural Katsina. It is imperative to note that vigilante activities has become glaring in the role they played over the years in peasant and pastoralists clash in different parts of Katsina in areas such as Mairuwa, Yankara, Ruwan Godia, Pauwa, Ketare and *Yar'kuka* villages in southern parts of the state (IDI, Male, Age 68, Kankara, Urban, 2013).

While the state and populace supported vigilante activities during the period on one hand, on the other hand this development vindicated the popular perception on state weakness in terms of police corruption and institutional corruption in the country for their failure to curtail the rising crime rates of the period (Fourchard,2003:7). It is important at this juncture to stressed that the activities of vigilante groups of this nature is not only peculiar to Nigerian societies alone. There exist other vigilante activities in Africa and Asian societies that were established with either similar or different motive especially in the period under review. For example, *Pam swakarsa* (self help security force) vigilante was among the famous vigilante activities in Nusa Tenggara Barat (NTB) province of Indonesia, established to convert growing crimes in the area (Kristiansen, 2003:121).

While the *Pam swakarsa*, *Sungu-sungu* and the *Yanbanga* groups in rural Katsina are aim at fighting proscriptive activities in their various *locale*. The vigilante groups found in the Kwa-zulu Natal in South Africa were organized and mobilised for purely political gains. Examples of such vigilante are the vigilantes used by the *Inkatha* Freedom Party (IFP) in South Africa under the leadership of Chief Bhutelezi in KwaZulu-Natal. Their activities in Kwazulu Natal between 1991 and 1993 became prominent, because the vigilantes were used as an armed wing of the IFP in the area in countering the growing popularity of the ANC in the region (Kevin, 2011:184-186). Figure 1 represents a specimen of identification card produced and use by the *vigilante* in Faskari LGA in the state.

Figure 1. Plate identification card use by YanBanga
Source: Snaped During field work at Mairuwa District, Faskari LGA

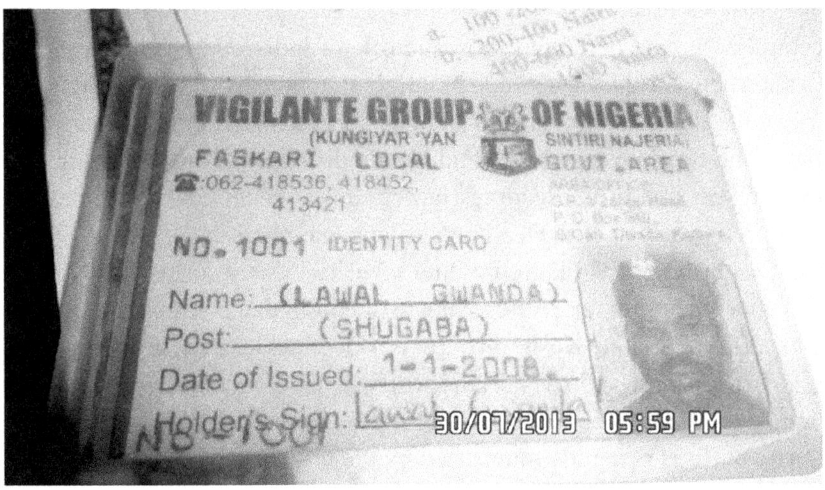

The vigilante activities during this period (1986-1999) can be contextualised on the basis of the economic hardship, corruption within state institutions. For example, corruption in Nigeria during the period became almost institutionalised phenomenon across public sectors. While there was the dwindling of the value of Naira, low agricultural output led to the deterioration of economy and rapid retrenchments of workers (in both private and public sectors of the economy) and social insecurity (Galadanci, 2001:42). On the other the formal structural changes in the state during the period as analysed based can be conceptulised as intentional structural changes that have greater impact on the society. The intentional economic changes by the state that have been adopted indicated the role of structure and human agency in reproduction of social practice. This is part of what Giddens described as intentional human agency that have unintended consequence as *Homeostatic loops*[1] (i.e., causal factors which have a feedback effect in system reproduction where that is largely the outcome of unintended consequences; Giddens, 2009:27b).

The state as an institution invented economic policies (the SAP's) is the intentional role human agency that have the vigilante activities as an unintended consequence as its feedback in the society. Thus, all the vigilante activities during the SAPs can be contextualised as an unintended consequence of the state actions because the state was not aware the *Kauraye* activities would be transformed. It further indicates changes in the economic setting of the area over time and space, which further influenced changes in the state economic structure and the social practice. The states structural properties here therefore became constrain and at the same enabling to the vigilante activities.

Adamu Koshe Phenomenon

The central argument here is to show the role of individual agents in the conflict reutilization between the farmers and pastoralist. Adamu Koshe came to the lime light because of his significant role in the fight against cattle rustlers that have bedeviled the communities of present Pauwa and Ketare districts of Katsina emirate. Adamu Koshe represent one of the leading figures in the recent conflict in the area. He was identified to be one of the leading vigilante members in the area. Vigilante and other self security outfits like represents security local security mostly composed of volunteers, operating under the mandate of communal consensus to fight crimes (Olaniyi, 2005). Adamu Koshe groups is was characterised by membership of different from within and outside the district, these includes friends, associates and immediate family members. As argued by Fourchad "the need for vigilante as security outfit is generally intensified by the growing state failure to meet local security needs, on one hand, and on the other hand, the growing desire for peace and rule of law among communities" (Abrahams, 1987; Fourchad, 2008). The existence of vigilante activities in the post-colonial period can be considered to be part of the people concern for both physical and spiritual insecurity that is being witnessed in the contemporary period (Last, 2016). In this context local security groups emerged as a result of the state's weak security apparatus, which resulted in the government's failure to fulfil its responsibility of providing maximum security to the public.

It is was part of his effort to combat the activities of fulani bandits in the area led to series of attacks to different fulani settlements in the area, this attacks can be described as outlaw activities in the area. Table 1 gives a summary of some of the fulani settlements attacked by the vigilante members across different village areas.

From Table 1, the settlements attacked by members of Koshe vigilante group is a clear manifestation of how volatile the district became during the period. These and many other fulani settlements were consistanly raided by the vigilante members, where houses and farm produce were completely destructed. Sources have indicated that apart from this the vigilante members develop the habit of imposing illegal fine to some of the fulani men they identified to be culprit to rustling. These actions of the vigilante at the same time forced many of the rustlers to abandon their family members and go into hiding in the moutain areas of Zango, Yar'tsamiyar Jino and neigbouring Zamfara state. The excessive activities of the vigilante forced many of the fulani pastorlist in the area boycot the attendance of weekly markets across the district. They did not only boycot the markets they at the same time sanctioned their women to stop supplying milk to all the markets in the district. They equally responded by attacks on some villages neighbouring the forest reserves. Villages such as Yar'tsamiyar Jino, Yar'santa and Bakai to mention but a few have suffered from these attacks by Fulani bandits. These attacks at the same time led to the relocation of many people from the villages to Kankara town (the district heaquaters) and other towns in the neibouring districts of DanMusa, Yantumaki and Yandaka. Among such that relocated were the village heads of Yar'santa and Yar'tsamiyar Jino. For

Table 1.

S/N	NAME OF FULANI SETTLEMENT	VILLAGE AREA	WARD
1	Gidan Alhaji Da'u	Kafin Runji	Sabon Gari
2	Gidan Shaffatu	Kafin Alhaji Burga	Sabon Gari
3	Gidan Jabbo	Danbirgima	Pauwa A & B
4	Gidan Sambo	Danbirgima	Pauwa A& B
5	Gidan Umar Sanda	Kwangerawa	Katoge
6	Gidan Isa Hete	Matsiga	Danmarabu
7	Gidan Sule Dare	Jeka Areda	Zango

example the village head of Yar'tsamiya was forced to relocate to Dutsin-ma, while that of Yar'santa relocated to Kankara town.

An important development in this respect was the public perception on the crisis in the area. Many people in the area for several years accused the security and the traditional institution of the area of convivence with both the notorious cattle rustlers and the farmers in perpetuating the crime against the innocent individuals in the area. Similarly, Onuoha (2015) argues that "the failure of public security in Nigeria has lead communities to indulge in different forms of 'self-help' security measures, ranging from vigilante groups to community-owned arms stockpiling" (Onuoha, 2015). The unprofessional attitudes and failure of security agencies in the fight against the cattle rustling bandits in the Pauwa region is clear in the events that followed the assassination of Adamu Koshe as the chief security officer in Katsina had to seek for immediate redeployment of security personnel in Kankara LGA (DailyTrust Newspaper, 2015)

Specifically vigilante activities were meant to combat crime when the government and the police had failed to effectively provide protection. To some extent it represents a sign of vote of no confidence in the state's ineffectiveness in providing public security to the general public. Before the incident of 10th September 2015 which led to the assassination of Adamu Koshe, many people in Nigeria are not aware of who Adamu Koshe was. The name came to known by many members of the public after his assassination by group notorious cattle rustling bandits that have identified him to a major threat to their activities in Southern Katsina. His assassination was at the same time followed by series of extra-judicial killing of many Fulanis that the vigilante members identified to be rustlers. Table 2 indicates some of the fulani rustlers that were killed by members of the vigilante in the area during the period.

Table 2.

S/N	Name of Victim	Area
1	Kofur Makau	Ketare
2	Isah Heiti	Danmarabu
3	Banyi	Kankara
4	Babai	Ketare
5	Ardo Kano	Kankara
6	Sani Riga	Yarsanta
7	Iroji	Kankara
8	Sambo	Pauwa A&B

Before his assassination many people across Nigeria many particularly in the areas ravaged by the ethno-religious crisis and Boko Haram insurgency troop to his place in Kankara LGA, Katsina state for the collection of charms. Koshe was known to a professional traditionalist who assist people with traditional charms that could protect them from gun shorts. People as far as Yalwan Shandam travelled to Kankara to collect charms which are mostly prepared on Sundays. Adamu Koshe is identified as a vigilante leader with many follows across many villages in Southern Katsina. Even though their was no comprehensive data on the exact number of his followers. His death and the events that followed after proved the number of followers he had. As reported by different daily paper upon his death "hundreds of members of the group brandishing various types of weapons barricaded the roads in protest over the killing (Leadership Newspaper).

The secretive attributes in form of charms by Koshe represents a strong ideological belief that surround the his activities and recognition. The attributes also indicates a strong traditional, social link and political link that exists between the group members and public, and within the circle members of vigilante. His group members and the society had considered these important focal practice that their activities revolves round within. Gwassa, indicates that explanations for philosophy or ideology over certain movement represents an important principle that is inspirational to certain violent activities (Gwassa,2005:86).

In searching for issues surrounding the possession of sprits and traditional medicine among the like groups in Kano, Casey identified people in the rural Kano to be more knowledgeable in ritual practices than the urban dwellers. He asserts that "...social networking that is maintained between different groups through the possession of superstitions and rituals powers associated with herbal medicines against sharp weapons is mostly the source of mutual bond between members" (Casey,2010:4). Thus, the local vigilante violence are associated with the services of traditional native doctors that provide voodoo and charms in form of herbal medicine for protection against weapons, their opposing the attacks by herdsmen who are equally armed with traditional locally made weapons.

CONCLUSION

By looking at the activities of vigilante in pauwa district, this paper has shown how the conflict between the farmers and the Fulani pastoralist led to emergence of banditory and vigilante activities in the area. The recent activities of different vigilante groups against the Fulani cattle rustlers in the Pauwa and other districts across Katsina emirate, highlight the role of traditional security arrangement in the fight against the pastoralist. Over the years as analysed the conflict between the farmers and Fulani pastoralist is characterised by number of socio-economic and political intricates. The failure of state policies particular the SAP's greatly influence the crisis. To understand the present form of vigilante activities provided by local militia group against cattle rustlers in different parts of Katsina Emirate, and the neigbouring emirates of Zamfara, Maru and Tsafe it is important to under score the role of individuals vigilante leaders the provision security in the area over the period. Traditional security method in form of vigilante remain part of the majority security apparatus in different parts of Hausaland. They are also typically a populist grass-roots development, although they may be used and abused by specific sectors of a local population. There is a strong element of self-help in them, but it is always self-help in the context of the state (Abrahams, 1987). Activities of vigilante during this period have depended on the role of both practical and discursive conscious agencies, through the possession of some charms that they obtained from soothsayers, and through their contacts with *jinn's.* They are believed to possess some magical powers. The belief in association with super natural powers and the possession of these particular charms by both the vigilante and Fulani pastoralist in Katsina society represents knowledgeability and cultural association. These however underscore the richness of Hausa traditional culture. It also represents the reasons why the vigilante are so much upheld and feared most in the society.

During the SAP period there was no qualitative change in the socio-economic sectors of the state, except continuity in the structures which were first produced during the colonial period. The Crisis continued to be under estimated as a result of weak leadership and bad policies through the institutionalisation of unemployment, poverty and growing corruption within the state. The present wave of conflicts across the different state of Nigerian federation may be attribute to quest of access to resources: hence the nagging problem of why large proportion of relatively literate Fulani always roam the forests which is considered to be a major reason behind existence of the phenomenon. This equally underscore the importance on emphasising on modern policies that will enhanced cattle grazing and modern farming techniques.

REFERENCES

Abrahams, R. (1987). Sungusungu: Village Vigilante Groups in Tanzania. *African Affairs*, *86*(343), 179-196. Retrieved from http://www.jstor.org/stable/722470

Adelakun, E. O., Adurogbangba, B., & Akinbile, L. A. (2015). Socioeconomic Effects of Farmer-Pastoralist Conflict on Agricultural Extension Service Delivery in Oyo State, Nigeria. *Journal of Agricultural Extension*, *19*(2), 59-70. Retrieved from http://journal.aesonnigeria.org

Aghedo, I., & Osumah, O. (2014). Bread, Not Bullets: Boko Haram and Insecurity Management in Northern Nigeria. *African Study Monographs*, *3-4*(35), 205-229.

AJCTC. (2015). *Aljazeerah Counting the Cost Program "Corruption Blights Nigerian Army's Fight Against Rebels*. Retrieved from http://www.aljazeera.com/programmes/countingthecost/2015/03/

Anderson, D. M. (2002). Vigilantes, Violence and the Politics of Public Order in Kenya. *African Affairs, 405*(101), 531-555. Retrieved from http://www.jstor.org/stable/3518466

Casey, C. (2014). *Hypocrisy, Spatial (In) Justice and Youthful Policing in Northern Nigeria*. Retrieved from http://books.openedition.org/pupo/436

Cohen, I. J. (2004). Structuration. In Encyclopedia of Social Theory (vol. 2). Sage Publications.

Federal Republic of Nigeria. (2009). Official 2006 Population Census Final Result. *Gazette*, *96*(2), 2009.

Fourchard, L. (2008). A New Name for an Old Practice: Vigilantes in South-Western Nigeria Africa. *Journal of the International African Institute*, *1*(78). Retrieved from http://www.jstor.org/stable/40027362.

Giddens, A. (1976). *New Rules of Sociological Method: A Positive Critique of Interpretative Sociologies*. Polity Press.

Giddens, A. (1990). *Central Problems in Social Theory: Action, Structure and Contradiction in Social Analysis*. University of California Press.

Giddens, A. (2009a). *The Constitution of Society: Outline of the Theory of Structuration*. Polity Press.

Giddens, A. (2009b). *The Consequence of Modernity*. Polity Press.

Gwassa, G. C. K. (2005). *The Outbreak and Development of the Maji Maji War 1905–1907*. Koln: Rudiger Koppe Verlag.

Hobsbawm, E. (2012). *Bandits*. London: Abacus.

Kofar Sauri, L. Y. (1986). *Pauwa in 19th and 20th Century* (B.A History Dissertation). Department of History, Bayero University, Kano.

Last, M. (2016). The Search for Security in Muslim Northern Nigeria. *Africa*, *78*(1).

Leadership Newspaper. (2013, November 19). Borno Trains Another 1,000. *Leadership Newspaper*.

Olaniyi, R. (2005). *Community Vigilantes in Metropolitan Kano 1985-2005* (Vol. 17). Ibadan: Institut Francais de Recherche en Afrique. doi:10.4000/books.ifra.727

Olayoku, P. A. (2014). *Trends and Patterns of Cattle Grazing and Rural Violence in Nigeria*. IFRA-Nigeria Working Papers Series, no 34. 28/11/2014.

Onuoha, F. C. (2015). Small Arms and Light Weapons Proliferation and Human Security in Nigeria. In *Conflict Trends*. African Centre for the Constructive Resolution of Disputes (ACCORD). Retrieved from www.accord.org.za

Oyama, S. (2014, October). Farmer-Herder Conflict, Land Rehabilitation, and Conflict Prevention in the Sahel Region of West Africa. *African Study Monographs*, (Suppl. 50), 103–122.

Salihi, H. M. (2012). Governance Failure and Political Violence in Nigeria: The Jama'atu Ahlils Sunah Lidda'awati wal-Jihad in Perspective. In Nigerian Convulsive Federalism: Perspectives on Flash-Points of Conflict in Northern Nigeria (pp. 211-233). Foludex Press Limited.

Saulawa, A. M. (1996). *The Colonial Factor: Migration in Birnin Katsina, c.1903-1949 A.D.* Zaria: Gaskiya Corporation Ltd.

Shuaibu, M. A. (2007). Traditional Military Recruitment and Weaponry in the Warfare of Hausaland 1500-1900. Kano Studies, 1(3), 53 – 68.

Weekly Trust Newspaper. (2014, September 6). *Boko Haram: Why Borno Residents are Arming Themselves*. Author.

Yari, L. (2007). *Muhammadu Dikko Emir of Katsina and His Life Times 1855-1944*. Kaduna: Prime Printers.

ENDNOTE

1 These are causal factors which have a feedback effect in system reproduction, where that is largely the outcome of unintended consequence.

Compilation of References

Aapengnuo, C. M. (2010). Misinterpreting ethnic conflicts in Africa, *Joint Force Quarterly, 58*(4), 13-17.

Abati, R. (2000). *The press, politics and society in Nigeria*. Lagos: Tosen Consult.

Abrahams, R. (1987). Sungusungu: Village Vigilante Groups in Tanzania. *African Affairs, 86*(343), 179-196. Retrieved from http://www.jstor.org/stable/722470

Abu-Fadil, M. (2017). *Was Intra Bank's Fall a Prelude to Lebanon's Downfall? Huffington Post*.

Abugu, S. O., & Onuba, C. O. (2015). Climate change and pastoral conflicts in the Middle Belt and south-east Nigeria: Implication on human resource of the regions. *Global Journal of Human Resource Management, 3*(5), 44–51.

Adelakun, E. O., Adurogbangba, B., & Akinbile, L. A. (2015). Socioeconomic Effects of Farmer-Pastoralist Conflict on Agricultural Extension Service Delivery in Oyo State, Nigeria. *Journal of Agricultural Extension, 19*(2), 59-70. Retrieved from http://journal.aesonnigeria.org

Adesiji, G. B., Akinsorotan, A. O., & Omokore, F. (2010). Farmers Assessment of extension services in Ogun state Nigeria. *Journal of Agricultural & Food Information, 11*(2), 143–156. doi:10.1080/10496501003691661

Adetola, B., & Johnson, B. (2017, October). Women rural farmers beg Federal Government to intervene in cattle herders' attack. Nigerian Tribune. *Tuesday, 17*, 7.

Adisa & Adekunle. (2010). Farmer-Herdsmen Conflicts: A factor analysis of socio-economic conflict variables among arable crop farmers in north central Nigeria. *Journal of Human Sociology., 30*(1), 1–9.

Aghedo, I., & Osumah, O. (2014). Bread, Not Bullets: Boko Haram and Insecurity Management in Northern Nigeria. *African Study Monographs, 3-4*(35), 205-229.

AJCTC. (2015). *Aljazeerah Counting the Cost Program "Corruption Blights Nigerian Army's Fight Against Rebels.* Retrieved from http://www.aljazeera.com/programmes/countingthecost/2015/03/

Akele, S. A., & Chukwu, G. O. (2004). *Poverty Alleviation through sustainable root and tuber crops production. In Proceeding of 8th triennial symposium-international society for tropical root crops, Africa Branch* (pp. 1–324). Ibadan, Nigeria: ISTRC AB.

Aker, F. (2014). *October 1973 - The Arab Israeli War* (1st ed.). Design Pub.

Akinfeleye, R. A. (2003). *Fourth estate of the realm or fouth estate of the wreck: Imperative of social responsibility of the press (inaugural lecture series).* Lagos: University of Lagos Press.

Alex, G. (2014). *Formative content.* Retrieved from http://www.google.com.ng/sea

Alimon, A. R. (2004). *The Nutritive Value of Palm Kernel Cake for Animal Feed. Palm Oil Development. No. 40, June 2004.* Malaysian Palm Oil Board.

Alubo, O. (2006). *Ethnic Conflicts and citizenship crises in the central region.* Ibadan: Programme On Ethnic and Federal Studies.

Alubo, S. O. (2006). *Ethnic conflicts and citizenship crises in the central region.* Ibadan: University of Ibadan.

Aluko, A. P., Oyeleye, B., Sulaiman, O. N., & Ukpe, I. E. (2008). Climate Change: A threat to Food security and environmental Protection. *Proceedings of the 32nd Annual Conference on Forestry Association of Nigeria.*

AMInfo. (2014). Middle Eastern clients in the HSBC Switzerland leaks. *Swiss Leaks.* Retrieved June 2017. Retrieved July 10, 2017, from http://ameinfo.com/luxury-lifestyle/list-middle-eastern-clients-in-the-hsbc-switzerland-leaks/

Anderson, B. (2005). Violent conflict: Bringing the Real threat into focus, Human Development Report. UNDP.

Anderson, D. M. (2002). Vigilantes, Violence and the Politics of Public Order in Kenya. *African Affairs, 405*(101), 531-555. Retrieved from http://www.jstor.org/stable/3518466

Anderson, J. (2001). *A popular Theatre/popular Education manual.* Toronto: Cartalyst Centre.

Ashworth, G. (1999). *The silence of women.* Cambridge, UK: CUP.

Asiedu, J. J. (1989). Processing Tropical Crops. ELBS/Macmillan.

Attoh, F., & Soyombo, O. (2011). The politics of ethnic balancing in Nigeria. *International Journal of Sociology and Anthropology, 3*(2), 40–44.

Awe, B. (2001). *Nigerian women: A historical perspective.* Ibadan: Bookcraft.

Awodiya, M. (1995). *The drama of Femi Osofisan: A critical perspective.* Ibadan: Kraft books.

Badejo, M. (1998). Agroecological restoration of savanna ecosystems. *Ecological Engineering, 10*(2), 209–219. doi:10.1016/S0925-8574(98)00012-3

Balasubramaniam, K. (1976). Polysaccharides of the kernel of maturing and mature coconuts. *Journal of Food Science, 41*(6), 1370–1373. doi:10.1111/j.1365-2621.1976.tb01174.x

Barkindo, A., Makanjuola, H., & Martins-Aginam, A. (2017). *2017 research report from the valley of death: Memory, healing and inter-group dialogue in southern Kaduna.* Abuja: Abbey Digital Prints.

Bedingar, T., & Degefa, G. (1990). *Trends in agro-byproducts and their feeding potential in sub-Saharan Africa.* Working document No. 14, ILCA, Addis-Ababa. Retrieved from http://www.fao.org/wairdocs/ILRI/x5480E/x5480E00.HTM

Best, S. G. (2007). *Conflict and peace building in Plateau State, Nigeria.* Ibadan: Spectrum.

Bitar, K., & de Fouchécour, C. (2015). *Le Cèdre et le Chêne. De Gaulle et le Liban – Les Libanais et de Gaulle.* Editions Geuthner. Retrieved May 3, 2016, from http://www.iris-france.org/publications/le-cedre-et-le-chene-de-gaulle-et-le-liban-les-libanais-et-de-gaulle/

Blake, G., & Wagstaff, M. (2016). *The Middle East: A Geographical Study* (2nd ed.). London, UK: RoutLedge.

Boal, A. (2000). *Theatre of the Oppressed.* New York: Theatre Communications Groups.

Boon, R., & Plashow, J. (2004). *Theatre and Empowerment: Community Drama on the world stage, case study from around the world*. University of Leeds. doi:10.1017/CBO9780511486166

Brecht, B. (1964). Brecht on Theatre. In J. Willett (Ed. & Trans.), The Development of an Aesthetics. London: Penthuen.

Brown, J. (2016). Wafic Said: businessman, philanthropist and political fixer. *Financial Times*. Retrieved May 15, 2017, from https://www.ft.com/content/a3cb764a-ecf1-11e5-bb79-2303682345c8

Caquot. (2017). Sémites. *EncyclopædiaUniversalis*. Retrieved May 18, 2017, fromhttp://www.universalis.fr/encyclopedie/semites/

Carpenter, C. (2002, Autumn). Gender Theory in World Politics: Contributions of a Non-Feminist Standpoint. *International Studies Review*, *4*(3), 153–165. doi:10.1111/1521-9488.00269

Carter, D. L., Thomas, R. J., & Ross, S. D. (2011). You are not a friend: Media conflict in times of peace. *Journalism Studies*, *12*(4), 456–473. doi:10.1080/1461670X.2010.530972

Casey, C. (2014). *Hypocrisy, Spatial (In) Justice and Youthful Policing in Northern Nigeria*. Retrieved from http://books.openedition.org/pupo/436

Chabal, P., & Daloz, J.-P. (1999). *Africa works: Disorder as political instrument*. Oxford, UK: International African Institute.

Chen, J., Lu, T., & Han, Z. (1997). Poultry production in China and the potential for using enzyme preparations. In *Enzymes in Poultry and Swine Nutrition* (pp. 154–155). IDRC Publication. http://www.idrc.ca/en/ev-30911-201-1-DO_TOPIC.html

Chinwoku, E. C. (2017). Glamorisation of Fulani herdsmen's criminality and its implications on national security in Nigeria. *Online Journal of Arts and Social Sciences*, *2*(2), 29–51.

Christian Aid. (2006). T*he Climate of poverty: facts and hope; A Christian Aid Report, May, 2006*. Author.

Clarke, Th., & Tigue, J. (1975). Dirty money: Swiss banks, the Mafia, money laundering, and white collar crime. Simon and Schuster.

Clark, J. P. (1962). *Ozidi*. London: Oxford University Press.

Cohen, I. J. (2004). Structuration. In Encyclopedia of Social Theory (vol. 2). Sage Publications.

Constable, A. L. (2017). Climate change and migration in the Pacific: Options for Tuvalu and the Marshall Islands. *Regional Environmental Change*, *17*(4), 1029–1038. doi:10.100710113-016-1004-5

D'Amato, G. (1995). Switzerland: A Multicultural Country without Multicultural Policies? In *The Multiculturalism Backlash European Discourses, Policies and Practices*. Swiss Forum for Migration and Population Studies, Neuchatel, Switzerland.

Daellenbach, H., McNickle, D., & Dye, Sh. (2012). *Management Science - Decision-making through systems thinking* (2nd ed.). Plagrave Macmillian.

Dahl, B. (2006). *The LebanesePalestinian Conflict in 1973: The Social (de) Construction of Lebanese Sovereignty*. St. Antony's College.

Danaan, G. N. (2017). A theoretical discourse on the mediatisation and framing of Jos ethno-religious conflicts. *Journal of Arts & Humanities*, *6*(2), 43–56. doi:10.18533/journal.v6i2.953

Danfulani, U. H. D. (2006). *The Jos peace conference and the indigene/settler question in Nigerian politics*. Paper presented at the Jos Peace Conference, Jos.

Deben, L. (2015, June 26). The Pope's critics are acting like children: The environmental encyclical makes us squirm because it expose our greed and recklessness with the powerful searching for the Gospel. *Catholic Herald,* 16-17.

Del Ponte, C., &Sudetic, C. (2009). *La traque, les criminels de guerre et moi: autobiographie*. ÉditionsHéloïsed'Ormesson.

Dennis, E., & John, C. M. (1984). *Basic issues in mass communication: A debate*. New York: Macmillan.

Dougherty, J.T., & Pfaltgraff. (1997). *Contending Theories of International Relation* (4th ed.). Addison Wesley, Education Publishers Inc.

Dung, P. S. (2005). *The Politicisation of Settler-Native Identities and Ethno-Religious Conflicts in Jos, Central Nigeria*. Stirling-Horden Publishers Ltd.

Dzeka, T. T. (2014). Ethnicity as a factor in Nigerian disunity. In V. Egwemi, T. Wuam & C. S. Orngu (Eds.), Federalism, Politics and Minorities in Nigeria: Essays in Honour of Professor G. N. Hembe (pp. 448–451). Lagos: Bahiti & Dalila Publishers.

Eagleton, T. (2012). *Literary theory: An introduction*. Minneapolis University Press.

Egwu, S. G. (2004). *Ethnicity and citizenship in urban Nigeria: The Jos case 1960-2000* (PhD Thesis). University of Jos, Jos.

Egwugwu, E, Steve, O. A., & Lantana, A. (1990). Article. *"NAWE" Journal of the Nigeria Popular Theatre Alliance, 2*(1), 108.

El Hashem, B. (1990). *It was Kissinger who destroyed the nation of Lebanon*. EIR Feature.

Elaigwu, I. J. (1994). *Ethnicity and Federal Option in Africa*. Abuja: National Council on Intergovernmental Relations.

El-Khazen, F. (2000). *The breakdown of the state in Lebanon, 1967-1976*. Harvard University Press.

Emenyonu, E. (Ed.). (2000). *Goatskin, bags and wisdom: New critical perspectives on African literature*. Trenton African World Press.

Enahoro, A. B. (2012). *Mitigating ICT effects on climate change in Nigeria*. Uyo: BSM Resources Limited.

Environment, F. M. o. (2015). *Nigeria's Intended Nationally Determined Contribution: Conference of Parties to the United Nations Framework Convention on Climate Change (COP-UNFCCC)*. Abuja: Federal Ministry of Environment.

FAO. (2007). Food and Agriculture organisation of the United Nations. In Climate change and Food security: a frame work document. FAO.

Faour, M. (2007). *Religion, demography, and politics in Lebanon*. Taylor & Francis.

Favier. (2017). *HaririRafic - (1944-2005)*. Encyclopædia Universalis. Retrieved June 10, 2017, from http://www.universalis.fr/encyclopedie/rafic-hariri/

Federal Republic of Nigeria. (2009). *Farmers led Approaches to Extension. Agricultural Research Extension Network. Abuja*. Overseas Development Institute.

Federal Republic of Nigeria. (2009). Official 2006 Population Census Final Result. *Gazette*, *96*(2), 2009.

Fersan, E. (2010). *Syro-Lebanese Migration (1880-Present): "Push" and "Pull" Factors*. Washington, DC: The Middle East Institute.

Forbes. (2016). *Edgar de Picciotto& family*. Retrieved July 6, 2017, from https://www.forbes.com/profile/edgar-de-picciotto/

Fourchard, L. (2008). A New Name for an Old Practice: Vigilantes in South-Western Nigeria Africa. *Journal of the International African Institute*, *1*(78). Retrieved from http://www.jstor.org/stable/40027362.

Frame, J. D. (2013). *Framing decision: Decision making that accounts for irrationality, people, and constraints*. San Francisco: Jossey-Bass.

Francis, A. (2002). *Riwa'iat*. Edition Nobilis. Liban.

Fregonese, S. (2009). The urbicide of Beirut? Geopolitics and the built environment in the Lebanese civil war (1975–1976). *Political Geography*, *28*(5), 309–318. doi:10.1016/j.polgeo.2009.07.005

Freire, P. (1970). *Pedagogy of the oppressed*. New York: Continuum.

Freire, P. (1999). *Pedagogy of the oppressed*. Penguin Books.

Frost, A., & Tarrow, R. (1990). *Improvisation in Drama*. Hampshire, UK: Macmillan.

Fwatshak, S. U. (2005-2006). Reconstructing the origins of the peoples of Plateau State: Questioning the 'we are all settlers' theory'. *Journal of historical society of Nigeria*, *1*(16), 122-140.

Galadima, J. D. (2010). *Brothers against brothers: The press, identity politics and conflict management in northern Nigeria*. Tripatide Services Limited.

Galadima, J. D., & Soola, E. O. (2012). The press and group identity politics in Nigeria: A study of newspaper framing of the 2003 and 2007 presidential elections. *Journal of Communication*, *10*(1), 1–27.

Galtung, J. (1992). *Global glasnost: Towards a new world information and communication order?* Cresskill, NJ: Hampton Press.

Gambo, B. (2010, March 7). Act now to avert Jol crisis escalations. *Sunday Standard.*

Gambo, B., & Dakop, K. (2010, January 31). Who wants Plateau on its knees? *The Sunday Standard.*

Genyi, G. A. (2014). *Ethnic and religious identities in shaping contestation for land based resources: The Tiv farmers and pastoralists conflicts in central Nigeria.* Paper presented at the Annual Conference on Ethnic and Religious Conflic Resolution and Peacebuilding, New York, NY.

Giddens, A. (1990). *Central Problems in Social Theory: Action, Structure and Contradiction in Social Analysis.* University of California Press.

Giddens, A. (1976). *New Rules of Sociological Method: A Positive Critique of Interpretative Sociologies.* Polity Press.

Giddens, A. (2009a). *The Constitution of Society: Outline of the Theory of Structuration.* Polity Press.

Giddens, A. (2009b). *The Consequence of Modernity.* Polity Press.

Gofwen, R. I. (2004). *Religious conflicts in northern Nigeria and nation buidling: The throes of two decades 1980-2000.* Kaduna: Human Rights Monitor.

Gofwen, R. I., & Ishaku, J. (2006). *State of emergency: Our stand by Plateau patriots.* Lagos: Bookhouse.

Gwassa, G. C. K. (2005). *The Outbreak and Development of the Maji Maji War 1905–1907.* Koln: Rudiger Koppe Verlag.

Habib, C. (2009). *Lebanese Politics and the Tyrany of Confessionalism.* Confluences Méditerranée. Retrieved June 11, 2017, from https://www.cairn.info/revue-confluences-mediterranee-2009-3-page-63.htm

Hackett, R. I. J. (2010). Managing or manipulating religious conflict in the Nigerian media. *Journal of Sustainable Development in Africa, 12*(3), 1–28.

Hair-Bejo, M., & Alimon, A. R. (1995). The protective role of zinc in palm kernel cake toxicity in sheep. *Malaysian Journal of Nutrition, 1,* 75–82. PMID:22692016

Haley, E., & Snider, L. (2006). *Lebanon in Crisis: Participants and Issues*. Syracuse University Press.

Hamel, I. (2003). Victoire à l'arraché d'un trafiquant d'armes. *Swissinfo*. Retrieved from https://www.swissinfo.ch/fre/victoire-%C3%A0-l-arrach%C3%A9-d-un-trafiquant-d-armes/3212952

Hanitzsch, T. (2007). Situating peace journalism in journalism studies: A critical appraisal. *Conflict and Communication*, *6*(2), 1–9.

Hartley, C. W. S. (1988). The Oil Palm (3rd ed.). Longman.

Hassan, O. A., & Yeong, S. W. (1999). By-products as animal feedstuffs. In S. Gurmit, K. H. Lim, L. Teo, & D. K. Lee (Eds.), *Oil Palm and The Environment: A Malaysian Perspective* (pp. 225–239). Malaysian Oil Palm Growers' Council.

Herman, E., & Chomsky, N. (1998). *Manufacturing consent: The political economy of the mass media*. New York: Pantheon Books.

Higazi, A. (2011). *The Jos crisis: A recurrent Nigerian tragedy*. Bonn: Friedrich-Ebert-Stiftung.

Hijazi, I. (1975). *Kissinger Has Nothing to Offer Assad, and Assad Has Nothing for Him*. The New York Times Archives. Retrieved January 13, 2016, from http://www.nytimes.com/1975/03/09/archives/kissinger-has-nothing-to-offer-assad-and-assad-has-nothing-for-him.html

Hitti, P. (2002). *History of the Arabs*. Palgrave. doi:10.1007/978-1-137-03982-8

Hobsbawm, E. (2012). *Bandits*. London: Abacus.

Hopquin, B. (2013). Attentat du Drakkar: qui a tué les paras français de Beyrouth en 1983? *Le Monde*. Retrieved August 7, 2017, from http://www.lemonde.fr/societe/article/2013/10/23/qui-a-tue-les-paras-francais-de-beyrouth-en-1983_3501317_3224.html#x1dP3v1VIc2M9yL6.99

Hutagalung, R. I., Mahyuddin, M., & Jalaludin, S. C. (1982). Feeds for farm animals from the oil palm. In The oil Palm in Agriculture in the Eighties (vol. 11). Incorporated Society of Planters.

Idire, H. E. (2007). *Economics of Demand and supply of vertinary services among commercial poultry farmers in Delta state* (Unpublished B. Agric. Thesis). Department of Agric. Economics, Delta State University.

Idris, H. (2010, April 20). Jos crises threat to national security-Miyetti Allah News. *Daily Trust*.

Imobighe, T. (2003). Ethnicity and Ethnic Conflicts in Nigeria: An Overview. In Civil Society and Ethnic Conflicts Management. Spectrum Books Ltd.

Indexmundi. (2018). Switzerland vs. Lebanon. *Indexmundi*. Retrieved February 13, 2018, from https://www.indexmundi.com/factbook/compare/switzerland.lebanon/economy

International Crisis Group. (2012). *Nigeria's indigene-settler divide*. London: Routledge.

Investopedia. (2017a). *Real Economic Growth Rate*. Investopedia. Retrieved March 22, 2017, fromhttp://www.investopedia.com/terms/r/realeconomicrate.asp

IPCC. (2008). *Climate Change and Water*. Available from: http://www.icc.ch/pdf/technical-papes/climate-change-water-en-pdf

Isard, W. (1992). *Understanding conflicts and the science of peace*. Blackwell.

Ishaku, J. (2012). *The road to Mogadishu: How Jihadist terrorism tears Nigeria apart*. Hamptul Press Ltd.

Ityavyar, D., & Obiajunwa, S. (1992). *The state and women in Nigeria*. Jos University Press.

Iyayi, E. A., & Davies, B. I. (2005). Effect of enzyme supplementation of palm kernel meal and brewer's dried grains on the performance of broilers. *International Journal of Poultry Science*, *4*(2), 76–80. doi:10.3923/ijps.2005.76.80

Jega, A. (2003). *Identity transformation and identity politics under structural adjustment in Nigeria*. Kano: Afrikainstitute/Apsalla/CRD.

Johnson, S. (1983). Francois Genoud: Terrorist controller for Swiss banks. *Executive Intelligence Review*.

Julian, H. (2014). *Lebanon Jews Denounce Israel as Beirut Synagogue Reopens*. Jewish Press. Retrieved July 10, 2017, from http://www.jewishpress.com/news/breaking-news/lebanon-jews-denounce-israel-as-beirut-synagogue-reopens/2014/05/20/

Kabaso, S. (2013). *Theatre for Development in Zambia*. Kabsy Digital Media.

Kaigama, I. A. (2012). *Peace, not war: A decade of interventions in the Plateau State crises (2001-2011)*. Hamtul Press Ltd.

Kalejauye, P. O., & Alliyu, N. (2013). Ethnic politics and social conflicts: Factors in Nigeria's underdevelopment. *Journal of International Social Research, 6*(27), 251–262.

Kant, I. (1946). *Despotism*. Retrieved from https://archive.org.details.despotis1946

Kant, I. (n.d.). *What is Enlightenment*. Available at http://www.columbia.edu/acis/ets/ ccread/etscc/kant.h

Katu, N. N. (2016). *Media, Conflict Audiencess and the Dynamics of Information Dissemination in Plateau State, Nigeria: Is the Tail Wagging the Dog?* (PhD Thesis). Ohio University, Athens, OH.

Kidd, R. (1980). People Theatre concretization and struggle in Media Development: Vol. 27. London: Macmillan.

Kofar Sauri, L. Y. (1986). *Pauwa in 19th and 20th Century* (B.A History Dissertation). Department of History, Bayero University, Kano.

Koss, R. (Ed.). (1997). *Aristotle's poetics*. New York: Dover Pub.

Kubursi, A. (1999). Reconstructing the Economy of Lebanon. Arab Studies Quarterly, 21(1).

L'Orient le Jour. (2017). Les juifs du Liban partent dans le sillage des guerres de 1967 et de 1975. *L'Orient le Jour.* Lebanon. Retrieved June 11, 2017, from https://www.lorientlejour.com/article/1066188/les-juifs-du-liban-partent-dans-le-sillage-des-guerres-de-1967-et-de-1975.html

Lar, I., & Embu, E. (2012). *Creative and Critical Writing for Peace Building in Nigeria: The Jos Paradigm*. Akin Press.

Last, M. (2016). The Search for Security in Muslim Northern Nigeria. *Africa, 78*(1).

Le Parisien. (2010). *History of Lebanon under Arab rule*. Le Parisien.

Leadership Newspaper. (2013, November 19). Borno Trains Another 1,000. *Leadership Newspaper*.

Lebanese Army. (2017). *Jean Njeim*. Lebanese Army. Retrieved May14, 2017, fromhttps://www.lebarmy.gov.lb/en/content/jean-njeim

Leed, K. T. (2006). *Conflict resolution theory and practice: Integration and application*. Manchester, UK: Manchester University Press.

Little, D. (2007). *His Finest Hour? Eisenhower, Lebanon, and the 1958 Middle East Crisis*. Oxford, UK: University Press.

Lohor, S. A. (2012). Pre-Colonial warefare among the Chadic-speakers of Jos Plateau and their neighbours (1800-1960). In O. Adeshina, O. Ogen, & N. E. Attah (Eds.), Critical Perspectives on Peace, Coflict and Warfare in Africa: Festschrift in honour of Siyan Oyeweso (pp. 177-186). Ile-Efe: Obafemi Awolowo Press.

Lynch, J. (2008). *Debates in peace journalism*. Sydney: Sydney University Press.

Lynch, J., & MacGoldrick, A. (2005). *Peace journalism*. Stroud: Hawthorn Press.

Macaulay, B. M. (2014). Land degradation in northern Nigeria: The impact and implications of human-related and climate factors. *African Journal of Environmental Science and Technology*, *8*(5), 267–273. doi:10.5897/AJEST2013.1584

Mangvwat, M. Y. (2013). *A history of class formation in the Plateau province of Nigeria, 1902-1960: The genesis of a ruling class*. Durham, NC: Carolina Academic Press.

Marino, E., & Ribot, J. (2012). Adding insult to injury: Climate change and the inequities of climate intervention. *Global Environmental Change*, *22*(2), 323–328. doi:10.1016/j.gloenvcha.2012.03.001

Matawal, A. B. (2012). The Jos Crises and National Security: A Historical Perspective. In *Creative and Critical Writing for Peace Building in Nigeria: The Jos Paradigm*. Akin Press.

Maurisse, M. (2016). *L'enfer des expatriés français en Suisse: une «enquête»*. Suisse: Le Temps.

Mazrui, A. A. (2000). Islam between ethnicity and economics: The dialectics of Africa's experience. In T. Salter & K. King (Eds.), Africa, Islam and Development: Islam and Development in Africa-African Islam, African Development (pp. 15–54). Edinburgh, UK: Centre of African Studies, University of Edinburgh.

Mbiti, J. S. (1963). *African religions and philosophy*. London: Heinemann Educational Publisher.

Microsoft Encarta, . (2008). [Computer software]. Redmond, WA: Microsoft Corporation.

Miller, J. (2017). Swiss high court rules anti-immigration SVP ad broke racism laws. *Reuters*. Retrieved June 11, 2017, from http://www.reuters.com/article/us-swiss-racism-svp-idUSKBN17F1UT

Miller, M. (2014). *Momentary Memorials: Political Posters of the Lebanese Civil War*. University of Colorado.

Mitchell, J. P. (2007). *Media violence and Christian ethics*. Cambridge, UK: Cambridge University Press.

Mohammed, G., Jaliye, M. M., Usman, S., & Baniwa, E. (2011). Adaptive Strategies to climate change variability and change in rural communities. *Proceedings of the 16th Annual National Conference of Agricultural Extension of Nigeria (AESON).*

Musa, A. O., & Ferguson, N. (2013). Enemy framing and the politics of reporting religious conflicts in the Nigerian press. *Media. War & Conflict*, *6*(1), 7–20. doi:10.1177/1750635212469909

Nnoli, O. (1978). *Ethnic politics in Nigeria*. Enugu: Fourth Dimension Publishing Company.

Nwapa, F. (2010). Women and creative writing in Africa. In *Landmarks and new essays in literary and linguistic theories and analysis.* Princeton.

Nwokolo, E. N., Bragg, D. B., & Saben, H. S. (1976). The availability of amino acids from palm kernel, soybean, cotton seed and rape seed meal for the growing chick. *Poultry Science*, *55*(6), 2300–2304. doi:10.3382/ps.0552300

Nwosu, L. E. (2012). Climate change and food security: Implication for effective agricultural extension services in Nigeria. *Proceedings of the 17th Annual National Conference of Agricultural Extension of Nigeria (AESON).*

O'Mara, F. P., Muligan, F. J., Cronin, E. J., Rath, M., & Caffrey, P. J. (1999). The nutritive value of palm kernel meal measured in vivo and using rumen fluid and enzymatic techniques. *Livestock Production Science*, *60*(2-3), 305–316. doi:10.1016/S0301-6226(99)00102-5

Obioha, E. E. (2009). Climate variability, Environmental change and Food Security Nexus in Nigeria. *Journal of Human Ecology (Delhi, India)*, *26*(2), 107–120. doi:10.1080/09709274.2009.11906172

Ochonu, M. (2008). Colonialism within colonialism: The Hausa-Caliphate imaginary and the British colonial administration of the Nigerian Middle Belt. *African Studies Quarterly*, *10*(2-3), 95–127.

Ofuya, C. O., & Nwajiuba, C. J. (1990). Microbial degradation and utilization of cassava peel. *World Journal of Microbiology & Biotechnology*, *6*(2), 144–146. doi:10.1007/BF01200933 PMID:24429985

Ogun State Diary. (2017). *Ogun State land, climate population*. Author.

Okai, D. B., & Opoku-Mensah, D. (1988). The effects of varying levels of palm kernel meal on growth performance and carcass characteristics of pigs. *Proc. 18-19th Ghana Animal Science Association Symposium*, 54-57.

Okai, D. B., Rhule, S. W. A., Nyannor, E. K. D., & Tandoh, I. (2006). *Feed Packages Based on Local Agro-industrial By-products for Small-scale Pig Farmers in the Ashanti Region*. Paper presented at the Project Leaders Meeting, AgSSIP Non-Ruminant Programme held at Animal Research Institute, Frafraha, Accra.

Okai, D. B., Olympio, O. S., & Anim, P. K. (2000). Responses of grower-finisher pigs to diets containing varying levels of wheat bran with or without Optizyme – an exogenous enzyme complex. *Journal of the University of Science and Technology*, *20*(1-3), 55–61.

Okeudo, N. J., Eboh, K. V., & Izugboekwe, N. V. (2005). Growth rate, carcass characteristics and organoleptic quality of broilers fed graded levels of palm kernel cake. *International Journal of Poultry Science*, *4*(5), 330–333. doi:10.3923/ijps.2005.330.333

Okidu, O. (2011). One state, many nations: Media portrayal of multiple identities in Nigeria. In L. Oso & U. A. Pate (Eds.), Mass Media and Society in Nigeria (pp. 49-62) Lagos: Malthouse Press Limited.

Okoli, I. C., Anyaegbunam, C. N., Etok, E. B., Uchegbo, M. C., & Udedibie, A. B. I. (2004). Socio- economic characteristics of poultry business entrepreneurs, Imo state, Nigeria. *Journal of Agriculture and Social Research*, *4*(2), 1–11.

Okonkwo, N. (Director). (2002). *Scorpion god* [Motion picture]. Nigeria: N.A.

Olaniyi, R. (2005). *Community Vigilantes in Metropolitan Kano 1985-2005* (Vol. 17). Ibadan: Institut Francais de Recherche en Afrique. doi:10.4000/books.ifra.727

Olayoku, P. A. (2014). *Trends and Patterns of Cattle Grazing and Rural Violence in Nigeria*. IFRA-Nigeria Working Papers Series, no 34. 28/11/2014.

Onakomaiyat, D. S. (2015). A Theoretical Extension of the Technology Acceptance Model: Four Longitudinal Fieldss Studies. *Journal of Management Sciences.*, *46*(2), 16–24.

Onuoha, F. C. (2015). Small Arms and Light Weapons Proliferation and Human Security in Nigeria. In *Conflict Trends*. African Centre for the Constructive Resolution of Disputes (ACCORD). Retrieved from www.accord.org.za

Onwubiko, K. (1985). *History of West Africa: School certificate. 1800-present day*. Onitsha, Anambra: Africana-FEP.

Osei, S. A., & Amo, J. (1987). Palm kernel cake as a broiler feed ingredient. *Poultry Science*, *66*(11), 1870–1873. doi:10.3382/ps.0661870 PMID:3447144

Oso, L. (2012). Mass media and democracy in Nigeria: The prospect of a Pan-Nigeria public sphere. In M. Mboho & H. Batta (Eds.), The Companion to Communication and Development Issues: Essays in Honour of Prof. Des Wilson (pp. 271-296). Uyo: BSM Resources Limited.

Osofisan, F. (1982). *Morountodun and other plays*. Ibadan: Longman.

Osofisan, F. (2006). *Women of Owu*. Ibadan: University Press.

Oteh, P. J. (2008). Our House: A Post-Colonial Performance Of Colonies. In *Crisis, in A Window on Africa: Changing Culture and Peoples*. Lagos, Nigeria: Dat and Partners.

Oyama, S. (2014, October). Farmer-Herder Conflict, Land Rehabilitation, and Conflict Prevention in the Sahel Region of West Africa. *African Study Monographs*, (Suppl. 50), 103–122.

Parker, S. (2016). *The demise of secret bank accounts – Switzerland's private banks. face a new era of transparency*. Wolters Kluwer Financial Services, Inc.

Peterson, S. (2011). *Why it Worked: Critical Success Factors of a Financial Reform Project in Africa*. Faculty Research Working Paper Series. Harvard Kennedy School.

Pimentel, da Silva, Rodrigues Silva, Resende Schio, de Olveira Rodrigues, & de Oliveira. (2015). *Feeding behaviour of lactating cows fed palm kernel cake in the diet*. Retrieved from www. Scielo.br/pdf/asas/37n1

Piscator, E. (1966). *The Piscator experiment*. Carbondale, IL: Southern Illinois University Press.

Pittock, B. A. ((2009). Climate Change: The Science, Impacts and Solution (2nd ed.). CSIRO Publishing.

Plateau Indigenous Development Association Network. (2010). *The history, ownership, establishment of Jos and the misconceptions about the recurrent Jos conflicts*. Jos: DAN-SiL.

Plateau State Government. (2004). *Plateau State of Nigeria: Government white paper on the report of the commission of inquiry into the riots of 12th April, 1994 in Jos metropolis September, 2004*. Jos: Government Printing Press.

Plateau State Government. (2009). *Government White Paper Report of the Judicial Commission of Inquiry into the civil disturbances in Jos and its environ, September 2001*. Jos: Ministry of Information and Communication.

Plateau State of Nigeria. (1997). Government Views and Decisions on the Report of the Administrative Committee on Gyero Communal Clashes in Jos South Local Government Area. Ministry of Information and Communication.

Plotnicov, L. (1972). Who owns Jos? Ethnic ideology in Nigerian urban politics. *Urban Anthropology, 1*(1), 1-13.

Rasaq, A. M. (2012). Conflict sensitive journalism and newspaper coverage of Jos conflict. *Entrepreneurial Journal of Management Sciences, 1*(1), 174–188.

Rhule, S. W. A. (1996). Growth rate and carcass characteristics of pigs fed on diets containing palm kernel cake. *Animal Feed Science and Technology, 61*(1-4), 167–172. doi:10.1016/0377-8401(95)00934-5

Rhule, S. W. A. (1998). The influence of type of palm kernel cake on the growth rate and carcass characteristics of pigs [Abstract]. *Ghana Journal of Agricultural Science, 31*(2). doi:10.4314/gjas.v31i2.1928

Rickman, R. (1999). *Swiss Banks and Jewish Souls by Gregg J. Rickman*. Central European History. JSTOR.

Rolland, J. (2004). Lebanon: Current Issues and Background. NOVA.

Romeo, L. (2011). *Charles Malek. Les Cles du Moyen Orient.* Retrieved August18, 2017, from http://www.lesclesdumoyenorient.com/Charles-Habib-Malek.html

Rotimi, O. (1972). *Kurunmi*. Ibadan: University Press.

Sachs, J. (2008). *Comon Wealth: Economics for a Crowed Planet*. London: Penguin Group.

Salami, I. (2001). *Emotan.* Jos: Mazlink.

Salami, I. (2004). *Idia, the warrior Queen*. Jos: Sanitez Pub.

Salawu, A. (2009). Media narrative construction of ethno-religious conflicts in Nigeria. *Estudos em Communicacao, 1*(5), 75–93.

Salawu, B., & Hassan, A. O. (2011). Ethnic politics and its implication. *Journal of Public Administration and Policy Research, 3*(2), 28–33.

Salibi, K. S. (1976). *Cross Roads to civil war: Lebanon, 1958-1976*. Delmar, NY: Caravan Books.

Salihi, H. M. (2012). Governance Failure and Political Violence in Nigeria: The Jama'atu Ahlils Sunah Lidda'awati wal-Jihad in Perspective. In Nigerian Convulsive Federalism: Perspectives on Flash-Points of Conflict in Northern Nigeria (pp. 211-233). Foludex Press Limited.

Samuel, O. O. (2012). *A historical account of ethnic conflicts in Nigeria*. GRIN Verlag GmbH.

Saulawa, A. M. (1996). *The Colonial Factor: Migration in Birnin Katsina, c.1903-1949 A.D.* Zaria: Gaskiya Corporation Ltd.

Scores Killed in Fresh Attacks in Plateau. (n.d.). Retrieved from http://saharareporters.com /2018/03/13

Seitz, J. L. (2008). *Global Issues: An Introduction* (3rd ed.). Oxford, UK: Blackwell Publishing.

Sen, A. (1999). *Development as freedom*. Oxford, UK: Oxford University Press.

Shaw, I. S. (2012). *Human Rights Journalism: Advances in reporting distant humanitarian interventions*. London: Palgrave. doi:10.1057/9780230358874

Shedrack, G. B. (2008). *Causes and Effects of Conflicts in the Southern Zone of Plateau State Nigeria*. Centre for Conflict Management and Peace Studies, University of Jos.

Shehu, B. S. (1997). Film, Culture and Democracy in Nigeria. In Y. Nasidi & I. Igoli (Eds.), *Culture and democracy*. Zaria: ABU Limited Press.

Shinar, D. (2003). The peace process in cultural conflict: The role of the media. *Conflict & Communication*, *2*(1), 1–10.

Shuaibu, M. A. (2007). Traditional Military Recruitment and Weaponry in the Warfare of Hausaland 1500-1900. Kano Studies, 1(3), 53 – 68.

Shukla, A. K., & Futrell, J. H. (1993). Collisional activation and dissociation of polyatomic ions. *Mass Spectrometry Reviews*, *12*(4), 211–255. doi:10.1002/mas.1280120402

Solomon, Ch. (2017). *The Ghost of AntounSaadeh Looms Large Over Syria.* WorldPolitics. Retrieved June 17, 2017, from https://raddingtonreport.com/ghost-antoun-saadeh-looms-large-syria/

Sorby, K. (2012). *Lebanon and the 1969 Cairo Agreement.* Archive. ProQuest.

Soyinka, W. (1976). *Myth literature and the African world.* Cambridge University Press.

Steiner, H. J., & Alston, P. (2000). *International human rights in context: Law, politics, morals: Texts and materials* (2nd ed.). London: Oxford University.

Sule, N. R. (2015). *Ethno-religious conflicts, mass media and national development: The Northern Nigeria experience* (PhD Thesis). University of Jos, Jos.

Sundu, B., & Dingle, J. (2003). Use of enzymes to improve the nutritional value of palm kernel meal and copra meal. *Proceedings of Queensland Poultry Science Symposium, 1*(14), 1-15.

Sundu, B., Kumar, A., & Dingle, J. (2006). Palm kernel meal in broiler diets: Effect on chicken performance and health. *World's Poultry Science Journal, 62*(02), 316–325. doi:10.1079/WPS2005100

Swami, V. (1922). Complete Works of Wami Vivekananda (Vol. 9). Vedanta Press.

Tabar, P. (2010). *The Lebanese Jewish Community: Emigration and Diasporic Relations.* Institute for Migration Studies and Associate Professor of Sociology/ Anthropology - Lebanese American University. Retrieved June 17, 2017, from http://www.ulcm.org/wlcu-lebanese-heritage/news/lebanese-emigration/2013/07/08/the-lebanese-jewish-community-emigration-and-diasporic-relations(dr-paul-tabar)

Taleb, N. (2012). *Antifragile: things that gain from disorder.* Library of Congress.

Tarazi-Fawaz, L. (2014). *A Land of Aching Hearts: The Middle East in the Great War.* Harvard University.

The FactBook CIA. (2018a). *Lebanon.* Retrieved February 13, 2018, from https://www.cia.gov/library/publications/the-world-factbook/geos/le.html

The FactBook CIA. (2018b). *Switzerland.* Retrieved February 13, 2018, from https://www.cia.gov/library/publications/the-world-factbook/geos/sz.html

The International Consortium of Investigative Journalists. (2015). *Banking Giant HSBC Sheltered Murky Cash Linked to Dictators and Arms Dealers*. Retrieved from https://www.icij.org/project/swiss-leaks/banking-giant-hsbc-sheltered-murky-cash-linked-dictators-and-arms-dealers

The Sun Newspaper. (n.d.). Retrieved from sunnewsonline.com/killing-ty-danjuma-dhq-and-the-poor-people-of-taraba/amp/

Toufic, K. (2004). *A political economy of Lebanon, 1948-2002: the limits of laissez-faire*. Library of Congress Cataloging-in-Publication Data.

Traboulsi, F. (2007). *A history of modern Lebanon*. Pluto Press.

Trad, A., & Kalpić, D. (2016a). A *Transformation Framework Proposal for Managers in Business Innovation and Business Transformation Projects - The role of transformation managers in organisational engineering*. Chinese American Scholars Association Conference E-Leader, Austria.

Trad, A., & Kalpić, D. (2015). *The (e)Business Transformation Framework for (e)commerce Architecture-Modeling Projects. In Encyclopedia of E-Commerce Development, Implementation, and Management*. IGI Global.

Trad, A., & Kalpić, D. (2017a). *An Intelligent Neural Networks Micro Artefact Patterns' Based Organisation architecture Model*. IGI Global. (submitted)

Trad, A., & Kalpić, D. (2017b). *Framework and Enterprise Architecture Framework for Managers in Business Innovation-The role of accounting in automated business environments*. IGI Global. (submitted)

Trad, A., & Kalpić, D. (2017c). A Neural Networks Portable and Agnostic Implementation Environment for Business Transformation Projects- The Basic Structure. In *IEEE Conference on Computational Intelligence*. IEEE. 10.1109/CIVEMSA.2017.7995318

Trad, A., & Kalpić, D. (2017d). A Neural Networks Portable and Agnostic Implementation Environment for Business Transformation Projects- The Framework. In *IEEE Conference on Computational Intelligence*. IEEE. 10.1109/CIVEMSA.2017.7995319

Trading Economics. (2017a). *Lebanon - GDP Annual Growth Rate*. Trading Economics. Retrieved April 10, 2017, from http://www.tradingeconomics.com/lebanon/gdp-growth-annual

Trading Economics. (2017b). *Syria - GDP Annual Growth Rate*. Trading Economics. April 10, 2017, from http://www.tradingeconomics.com/syria/gdp-growth-annual

Trading Economics. (2017c). *Israel - GDP Annual Growth Rate*. Trading Economics. April 10, 2017, from http://www.tradingeconomics.com/israel/gdp-growth-annual

Trading Economics. (2017d). *France - GDP Annual Growth Rate*. Trading Economics. April 10, 2017, from http://www.tradingeconomics.com/france/gdp-growth-annual

Trading Economics. (2017e). *United States - GDP Annual Growth Rate*. Trading Economics. April 10, 2017, from http://www.tradingeconomics.com/united-states/gdp-growth-annual

Trading Economics. (2017f). *United Kingdom - GDP Annual Growth Rate*. Trading Economics. April 10, 2017, from http://www.tradingeconomics.com/united-kingdom/gdp-growth-annual

Trading Economics. (2017g). *Switzerland - GDP Annual Growth Rate*. Trading Economics. April 10, 2017, from http://www.tradingeconomics.com/switzerland/gdp-growth-annual

Trading Economics. (2017h). *Cyprus - GDP Annual Growth Rate*. Trading Economics. April 10, 2017, from http://www.tradingeconomics.com/cyprus/gdp-growth-annual

Trading Economics. (2017i). *Switzerland's currency evolution*. Trading Economics.

Trading Economics. (2017j). *Lebanon's currency evolution*. Trading Economics.

Tribune de Geneve. (2014). *Pour Christian Levrat, l'UDC est sur la voie du fascisme*. Tribune de Geneve. Retrieved July 12, 2017, from https://mobile2.tdg.ch/articles/551c50ecab5c3729aa000002

Trigiani, K. (n.d.). *Masculinity-Femininity: Society's Difference Dividend*. Retrieved from http://web2.airmail.net/ktrig246/out_of_cave/mf.html

Udoaka, N. (2015). Reporting ethnic minority issues in Africa: A study of Nigerian newspapers. In W. Mano (Ed.), *Racism, Ethnicity and the Media in Africa: Mediating Conflict in the Twenty-First Century* (pp. 117–136). London: I. B. Tauris.

Ugor, P. (2013). Nollywood and Post-colonial Predicaments: Transnationalism, Gender, and Commoditisation of Desire in Glamour Girls. In *Global Nollywood; The Transnational Dimentions of An African Video Film Industry*. Bloomington, IN: Indiana University Press.

Umar, M., & Ossai, E. (2014). Ethnicity and The Dilemma of Integration in Nigeria. *Current Issues in Development, 2*(3).

UN. (1992). *United Nations Framework Convention on Climate Change*. Available from: http:/unfcc.int/resource/does/conveng.pdf

UN. (n.d.). *Taif Accords*. UN. Retrieved April 10, 2017, from http://peacemaker.un.org/lebanon-taifaccords89

UNDP. (2007). Human Development Report 2007/2008: Climate change: Human solidarity in a divided world. Palgrave Macmillan.

Vloeberghs, W. (2015). *Architecture, Power and Religion in Lebanon: Rafiq Hariri and the Politics of Sacred Space in Beirut*. BRILL.

Weekly Trust Newspaper. (2014, September 6). *Boko Haram: Why Borno Residents are Arming Themselves*. Author.

Whetmore, J. E. (1985). *Media America* (3rd ed.). Wadsworth Publishing Company.

WHO. (2018). *Gender*. Retrieved January 9, 2015 from www.who.int/gender

Wolff, J. (1993). *The social production of art*. New York: Macmillan. doi:10.1007/978-1-349-23041-9

World Journals. (2002). *Petro Trad*. World Heritage Encyclopedia. Retrieved from http://worldjournals.org/articles/Petro_Trad

Yari, L. (2007). *Muhammadu Dikko Emir of Katsina and His Life Times 1855-1944*. Kaduna: Prime Printers.

Yeong, S. W. (1983). *Amino acids availability of palm kernel cake, palm oil sludge and sludge fermented product (Prolima) in studies with chickens*. MARDI.

Yusha'u, M. J. (2015). Regionalism and ethnicity in the Nigerian press: An analysis of the coverage of Boko Haram and the Niger Delta conflict in the Guardian and Daily Trust. In M. Wiston (Ed.), *Racism, Ethnicity and the Media in Africa: Mediating Conflict in the Twenty-First Century* (pp. 137–156). London: I.B. Tauris & Co Ltd.

Zalloua, P. (2004). *The NGM Study "Who were the Phoenicians" and the Return of the Phoenicians*. Interview Lebanese Broadcasting Corporation (LBC). Retrieved April 16, 2017, from http://phoenicia.org/genetics.html

Zartman, W. I. (1991). *Conflict resolution in Africa*. Washington, DC: The Brookings Institute.

Related References

To continue our tradition of advancing academic research, we have compiled a list of recommended IGI Global readings. These references will provide additional information and guidance to further enrich your knowledge and assist you with your own research and future publications.

Abresch, J. (2003). Geographic Information Systems Research and Data Centers. In A. Hanson & B. Levin (Eds.), *Building a Virtual Library* (pp. 52–64). Hershey, PA: IGI Global. doi:10.4018/978-1-59140-106-3.ch004

Ajiboye, O. E., & Yusuff, O. S. (2017). Foreign Land Acquisition: Food Security and Food Chains – The Nigerian Experience. In I. Management Association (Ed.), Natural Resources Management: Concepts, Methodologies, Tools, and Applications (pp. 1524-1545). Hershey, PA: IGI Global. doi:10.4018/978-1-5225-0803-8.ch072

Alapiki, H. E., & Amadi, L. A. (2018). Sustainable Food Consumption in the Neoliberal Order: Challenges and Policy Implications. In A. Obayelu (Ed.), *Food Systems Sustainability and Environmental Policies in Modern Economies* (pp. 90–123). Hershey, PA: IGI Global. doi:10.4018/978-1-5225-3631-4.ch005

Altaş, A. (2018). Geographical Information System Applications Utilized in Museums in Turkey Within the Scope of the Cultural Heritage Tourism: A Case Study of Mobile Application of Müze Asist. In S. Chaudhuri & N. Ray (Eds.), *GIS Applications in the Tourism and Hospitality Industry* (pp. 42–60). Hershey, PA: IGI Global. doi:10.4018/978-1-5225-5088-4.ch002

Andreea, I. R. (2018). Beyond Macroeconomics of Food and Nutrition Security. *International Journal of Sustainable Economies Management*, *7*(1), 13–22. doi:10.4018/IJSEM.2018010102

Anwar, J. (2017). Reproductive and Mental Health during Natural Disaster: Implications and Issues for Women in Developing Nations – A Case Example. In I. Management Association (Ed.), Gaming and Technology Addiction: Breakthroughs in Research and Practice (pp. 446-472). Hershey, PA: IGI Global. doi:10.4018/978-1-5225-0778-9.ch021

Awadh, H., Aksissou, M., Benhardouze, W., Darasi, F., & Snaiki, J. (2018). Socioeconomic Status of Artisanal Fishers in the West Part of Moroccan Mediterranean. *International Journal of Social Ecology and Sustainable Development*, *9*(1), 40–52. doi:10.4018/IJSESD.2018010104

Aye, G. C., & Haruna, R. F. (2018). Effect of Climate Change on Crop Productivity and Prices in Benue State, Nigeria: Implications for Food Security. In V. Erokhin (Ed.), *Establishing Food Security and Alternatives to International Trade in Emerging Economies* (pp. 244–268). Hershey, PA: IGI Global. doi:10.4018/978-1-5225-2733-6.ch012

Azizan, S. A., & Suki, N. M. (2017). Consumers' Intentions to Purchase Organic Food Products. In T. Esakki (Ed.), *Green Marketing and Environmental Responsibility in Modern Corporations* (pp. 68–82). Hershey, PA: IGI Global. doi:10.4018/978-1-5225-2331-4.ch005

Barakabitze, A. A., Fue, K. G., Kitindi, E. J., & Sanga, C. A. (2017). Developing a Framework for Next Generation Integrated Agro Food-Advisory Systems in Developing Countries. In I. Management Association (Ed.), Agri-Food Supply Chain Management: Breakthroughs in Research and Practice (pp. 47-67). Hershey, PA: IGI Global. doi:10.4018/978-1-5225-1629-3.ch004

Beachcroft-Shaw, H., & Ellis, D. (2018). Using Successful Cases to Promote Environmental Sustainability: A Social Marketing Approach. In I. Management Association (Ed.), Sustainable Development: Concepts, Methodologies, Tools, and Applications (pp. 936-953). Hershey, PA: IGI Global. doi:10.4018/978-1-5225-3817-2.ch042

Behnassi, M., Kahime, K., Boussaa, S., Boumezzough, A., & Messouli, M. (2017). Infectious Diseases and Climate Vulnerability in Morocco: Governance and Adaptation Options. In I. Management Association (Ed.), Public Health and Welfare: Concepts, Methodologies, Tools, and Applications (pp. 91-109). Hershey, PA: IGI Global. doi:10.4018/978-1-5225-1674-3.ch005

Bekele, F., & Bekele, I. (2017). Social and Environmental Impacts on Agricultural Development. In W. Ganpat, R. Dyer, & W. Isaac (Eds.), *Agricultural Development and Food Security in Developing Nations* (pp. 21–56). Hershey, PA: IGI Global. doi:10.4018/978-1-5225-0942-4.ch002

Benaouda, A., & García-Peñalvo, F. J. (2018). Towards an Intelligent System for the Territorial Planning: Agricultural Case. In F. García-Peñalvo (Ed.), *Global Implications of Emerging Technology Trends* (pp. 158–178). Hershey, PA: IGI Global. doi:10.4018/978-1-5225-4944-4.ch010

Bhaskar, A., Rao, G. B., & Vencatesan, J. (2017). Characterization and Management Concerns of Water Resources around Pallikaranai Marsh, South Chennai. In P. Rao & Y. Patil (Eds.), *Reconsidering the Impact of Climate Change on Global Water Supply, Use, and Management* (pp. 102–121). Hershey, PA: IGI Global. doi:10.4018/978-1-5225-1046-8.ch007

Bianchini, E., & Sicurella, S. (2010). GIS: A New Tool for Criminology and Victimology's Studies. In R. Sette (Ed.), *Cases on Technologies for Teaching Criminology and Victimology: Methodologies and Practices* (pp. 87–110). Hershey, PA: IGI Global. doi:10.4018/978-1-60566-872-7.ch006

Bogataj, D., & Drobne, D. (2017). Control of Perishable Goods in Cold Logistic Chains by Bionanosensors. In S. Joo (Ed.), *Applying Nanotechnology for Environmental Sustainability* (pp. 376–402). Hershey, PA: IGI Global. doi:10.4018/978-1-5225-0585-3.ch016

Bogataj, D., & Drobne, D. (2017). Control of Perishable Goods in Cold Logistic Chains by Bionanosensors. In I. Management Association (Ed.), Materials Science and Engineering: Concepts, Methodologies, Tools, and Applications (pp. 471-497). Hershey, PA: IGI Global. doi:10.4018/978-1-5225-1798-6.ch019

Bogueva, D., & Marinova, D. (2018). What Is More Important: Perception of Masculinity or Personal Health and the Environment? In D. Bogueva, D. Marinova, & T. Raphaely (Eds.), *Handbook of Research on Social Marketing and Its Influence on Animal Origin Food Product Consumption* (pp. 148–162). Hershey, PA: IGI Global. doi:10.4018/978-1-5225-4757-0.ch010

Bouzid, M. (2017). Waterborne Diseases and Climate Change: Impact and Implications. In M. Bouzid (Ed.), *Examining the Role of Environmental Change on Emerging Infectious Diseases and Pandemics* (pp. 89–108). Hershey, PA: IGI Global. doi:10.4018/978-1-5225-0553-2.ch004

Bowles, D. C. (2017). Climate Change-Associated Conflict and Infectious Disease. In M. Bouzid (Ed.), *Examining the Role of Environmental Change on Emerging Infectious Diseases and Pandemics* (pp. 68–88). Hershey, PA: IGI Global. doi:10.4018/978-1-5225-0553-2.ch003

Buck, J. J., & Lowry, R. K. (2017). Oceanographic Data Management: Quills and Free Text to the Digital Age and "Big Data". In P. Diviacco, A. Leadbetter, & H. Glaves (Eds.), *Oceanographic and Marine Cross-Domain Data Management for Sustainable Development* (pp. 1–22). Hershey, PA: IGI Global. doi:10.4018/978-1-5225-0700-0.ch001

Buse, C. G. (2017). Are Climate Change Adaptation Policies a Game Changer?: A Case Study of Perspectives from Public Health Officials in Ontario, Canada. In M. Bouzid (Ed.), *Examining the Role of Environmental Change on Emerging Infectious Diseases and Pandemics* (pp. 230–257). Hershey, PA: IGI Global. doi:10.4018/978-1-5225-0553-2.ch010

Calderon, F. A., Giolo, E. G., Frau, C. D., Rengel, M. G., Rodriguez, H., Tornello, M., ... Gallucci, R. (2018). Seismic Microzonation and Site Effects Detection Through Microtremors Measures: A Review. In N. Ceryan (Ed.), *Handbook of Research on Trends and Digital Advances in Engineering Geology* (pp. 326–349). Hershey, PA: IGI Global. doi:10.4018/978-1-5225-2709-1.ch009

Carfì, D., Donato, A., & Panuccio, D. (2018). A Game Theory Coopetitive Perspective for Sustainability of Global Feeding: Agreements Among Vegan and Non-Vegan Food Firms. In I. Management Association (Ed.), Game Theory: Breakthroughs in Research and Practice (pp. 71-104). Hershey, PA: IGI Global. doi:10.4018/978-1-5225-2594-3.ch004

Carr, D. A., & Carr, T. R. (2007). Geographic Information System Applications in the Public Sector. In G. Garson (Ed.), *Modern Public Information Technology Systems: Issues and Challenges* (pp. 293–311). Hershey, PA: IGI Global. doi:10.4018/978-1-59904-051-6.ch015

Castagnolo, V. (2018). Analyzing, Classifying, Safeguarding: Drawing for the Borgo Murattiano Neighbourhood of Bari. In G. Carlone, N. Martinelli, & F. Rotondo (Eds.), *Designing Grid Cities for Optimized Urban Development and Planning* (pp. 93–108). Hershey, PA: IGI Global. doi:10.4018/978-1-5225-3613-0.ch006

Chekima, B. (2018). The Dilemma of Purchase Intention: A Conceptual Framework for Understanding Actual Consumption of Organic Food. *International Journal of Sustainable Economies Management*, *7*(2), 1–13. doi:10.4018/IJSEM.2018040101

Chen, Y. (2017). Sustainable Supply Chains and International Soft Landings: A Case of Wetland Entrepreneurship. In B. Christiansen & F. Kasarcı (Eds.), *Corporate Espionage, Geopolitics, and Diplomacy Issues in International Business* (pp. 232–247). Hershey, PA: IGI Global. doi:10.4018/978-1-5225-1031-4.ch013

Çıtak, L., Akel, V., & Ersoy, E. (2018). Investors' Reactions to the Announcement of New Constituents of BIST Sustainability Index: An Analysis by Event Study and Mean-Median Tests. In M. Risso & S. Testarmata (Eds.), *Value Sharing for Sustainable and Inclusive Development* (pp. 270–289). Hershey, PA: IGI Global. doi:10.4018/978-1-5225-3147-0.ch012

Crossland, M. D. (2009). Geographic Information Systems as Decision Tools. In M. Khosrow-Pour, D.B.A. (Ed.), Encyclopedia of Information Science and Technology, Second Edition (pp. 1630-1633). Hershey, PA: IGI Global. doi:10.4018/978-1-60566-026-4.ch257

Crossland, M. D., Herschel, R. T., Perkins, W. C., & Scudder, J. N. (2002). Geographic Information Systems: How Cognitive Style Impacts Decision-Making Effectiveness. In M. Mahmood (Ed.), *Advanced Topics in End User Computing* (Vol. 1, pp. 285–301). Hershey, PA: IGI Global. doi:10.4018/978-1-930708-42-6.ch017

D'Aleo, V., D'Aleo, F., & Bonanno, R. (2018). New Food Industries Toward a New Level of Sustainable Supply: Success Stories, Business Models, and Strategies. In V. Erokhin (Ed.), *Establishing Food Security and Alternatives to International Trade in Emerging Economies* (pp. 74–97). Hershey, PA: IGI Global. doi:10.4018/978-1-5225-2733-6.ch004

Dagevos, H., & Reinders, M. J. (2018). Flexitarianism and Social Marketing: Reflections on Eating Meat in Moderation. In D. Bogueva, D. Marinova, & T. Raphaely (Eds.), *Handbook of Research on Social Marketing and Its Influence on Animal Origin Food Product Consumption* (pp. 105–120). Hershey, PA: IGI Global. doi:10.4018/978-1-5225-4757-0.ch007

Deenapanray, P. N., & Ramma, I. (2017). Adaptations to Climate Change and Climate Variability in the Agriculture Sector in Mauritius: Lessons from a Technical Needs Assessment. In I. Management Association (Ed.), Natural Resources Management: Concepts, Methodologies, Tools, and Applications (pp. 655-680). Hershey, PA: IGI Global. doi:10.4018/978-1-5225-0803-8.ch030

Deenapanray, P. N., & Ramma, I. (2017). Adaptations to Climate Change and Climate Variability in the Agriculture Sector in Mauritius: Lessons from a Technical Needs Assessment. In I. Management Association (Ed.), Natural Resources Management: Concepts, Methodologies, Tools, and Applications (pp. 655-680). Hershey, PA: IGI Global. doi:10.4018/978-1-5225-0803-8.ch030

Deshpande, S., Basu, S. K., Li, X., & Chen, X. (2017). Smart, Innovative and Intelligent Technologies Used in Drug Designing. In I. Management Association (Ed.), Pharmaceutical Sciences: Breakthroughs in Research and Practice (pp. 1175-1191). Hershey, PA: IGI Global. doi:10.4018/978-1-5225-1762-7.ch045

Dlamini, P. N. (2017). Use of Information and Communication Technologies Tools to Capture, Store, and Disseminate Indigenous Knowledge: A Literature Review. In P. Ngulube (Ed.), *Handbook of Research on Theoretical Perspectives on Indigenous Knowledge Systems in Developing Countries* (pp. 225–247). Hershey, PA: IGI Global. doi:10.4018/978-1-5225-0833-5.ch010

Dolejsova, M., & Kera, D. (2017). The Fermentation GutHub Project and the Internet of Microbes. In S. Konomi & G. Roussos (Eds.), *Enriching Urban Spaces with Ambient Computing, the Internet of Things, and Smart City Design* (pp. 25–46). Hershey, PA: IGI Global. doi:10.4018/978-1-5225-0827-4.ch002

Dolunay, O. (2018). A Paradigm Shift: Empowering Farmers to Eliminate the Waste in the Form of Fresh Water and Energy Through the Implementation of 4R+T. In I. Management Association (Ed.), Sustainable Development: Concepts, Methodologies, Tools, and Applications (pp. 882-892). Hershey, PA: IGI Global. doi:10.4018/978-1-5225-3817-2.ch039

Dube, P., Heijman, W. J., Ihle, R., & Ochieng, J. (2018). The Potential of Traditional Leafy Vegetables for Improving Food Security in Africa. In V. Erokhin (Ed.), *Establishing Food Security and Alternatives to International Trade in Emerging Economies* (pp. 220–243). Hershey, PA: IGI Global. doi:10.4018/978-1-5225-2733-6.ch011

Duruji, M. M., & Urenma, D. F. (2017). The Environmentalism and Politics of Climate Change: A Study of the Process of Global Convergence through UNFCCC Conferences. In I. Management Association (Ed.), *Natural Resources Management: Concepts, Methodologies, Tools, and Applications* (pp. 77-108). Hershey, PA: IGI Global. 10.4018/978-1-5225-0803-8.ch004

Dutta, U. (2017). Agro-Geoinformatics, Potato Cultivation, and Climate Change. In S. Londhe (Ed.), *Sustainable Potato Production and the Impact of Climate Change* (pp. 247–271). Hershey, PA: IGI Global. doi:10.4018/978-1-5225-1715-3.ch012

Edirisinghe, R., Stranieri, A., & Wickramasinghe, N. (2017). A Taxonomy for mHealth. In N. Wickramasinghe (Ed.), *Handbook of Research on Healthcare Administration and Management* (pp. 596–615). Hershey, PA: IGI Global. doi:10.4018/978-1-5225-0920-2.ch036

Ekpeni, N. M., & Ayeni, A. O. (2018). Global Natural Hazard and Disaster Vulnerability Management. In A. Eneanya (Ed.), *Handbook of Research on Environmental Policies for Emergency Management and Public Safety* (pp. 83–104). Hershey, PA: IGI Global. doi:10.4018/978-1-5225-3194-4.ch005

Ene, C., Voica, M. C., & Panait, M. (2017). Green Investments and Food Security: Opportunities and Future Directions in the Context of Sustainable Development. In M. Mieila (Ed.), *Measuring Sustainable Development and Green Investments in Contemporary Economies* (pp. 163–200). Hershey, PA: IGI Global. doi:10.4018/978-1-5225-2081-8.ch007

Escamilla, I., Ruíz, M. T., Ibarra, M. M., Soto, V. L., Quintero, R., & Guzmán, G. (2018). Geocoding Tweets Based on Semantic Web and Ontologies. In M. Lytras, N. Aljohani, E. Damiani, & K. Chui (Eds.), *Innovations, Developments, and Applications of Semantic Web and Information Systems* (pp. 372–392). Hershey, PA: IGI Global. doi:10.4018/978-1-5225-5042-6.ch014

Escribano, A. J. (2018). Marketing Strategies for Trendy Animal Products: Sustainability as a Core. In F. Quoquab, R. Thurasamy, & J. Mohammad (Eds.), *Driving Green Consumerism Through Strategic Sustainability Marketing* (pp. 169–203). Hershey, PA: IGI Global. doi:10.4018/978-1-5225-2912-5.ch010

Eudoxie, G., & Roopnarine, R. (2017). Climate Change Adaptation and Disaster Risk Management in the Caribbean. In W. Ganpat & W. Isaac (Eds.), *Environmental Sustainability and Climate Change Adaptation Strategies* (pp. 97–125). Hershey, PA: IGI Global. doi:10.4018/978-1-5225-1607-1.ch004

Farmer, L. S. (2017). Data Analytics for Strategic Management: Getting the Right Data. In V. Wang (Ed.), *Encyclopedia of Strategic Leadership and Management* (pp. 810–822). Hershey, PA: IGI Global. doi:10.4018/978-1-5225-1049-9.ch056

Fattal, L. R. (2017). Catastrophe: An Uncanny Catalyst for Creativity. In R. Shin (Ed.), *Convergence of Contemporary Art, Visual Culture, and Global Civic Engagement* (pp. 244–262). Hershey, PA: IGI Global. doi:10.4018/978-1-5225-1665-1.ch014

Forti, I. (2017). A Cross Reading of Landscape through Digital Landscape Models: The Case of Southern Garda. In A. Ippolito (Ed.), *Handbook of Research on Emerging Technologies for Architectural and Archaeological Heritage* (pp. 532–561). Hershey, PA: IGI Global. doi:10.4018/978-1-5225-0675-1.ch018

Gharbi, A., De Runz, C., & Akdag, H. (2017). Urban Development Modelling: A Survey. In S. Faiz & K. Mahmoudi (Eds.), *Handbook of Research on Geographic Information Systems Applications and Advancements* (pp. 96–124). Hershey, PA: IGI Global. doi:10.4018/978-1-5225-0937-0.ch004

Ghosh, I., & Ghoshal, I. (2018). Implications of Trade Liberalization for Food Security Under the ASEAN-India Strategic Partnership: A Gravity Model Approach. In V. Erokhin (Ed.), *Establishing Food Security and Alternatives to International Trade in Emerging Economies* (pp. 98–118). Hershey, PA: IGI Global. doi:10.4018/978-1-5225-2733-6.ch005

Glaves, H. M. (2017). Developing a Common Global Framework for Marine Data Management. In P. Diviacco, A. Leadbetter, & H. Glaves (Eds.), *Oceanographic and Marine Cross-Domain Data Management for Sustainable Development* (pp. 47–68). Hershey, PA: IGI Global. doi:10.4018/978-1-5225-0700-0.ch003

Godulla, A., & Wolf, C. (2018). Future of Food: Transmedia Strategies of National Geographic. In R. Gambarato & G. Alzamora (Eds.), *Exploring Transmedia Journalism in the Digital Age* (pp. 162–182). Hershey, PA: IGI Global. doi:10.4018/978-1-5225-3781-6.ch010

Gomes, P. P. (2018). Food and Environment: A Review on the Sustainability of Six Different Dietary Patterns. In A. Obayelu (Ed.), *Food Systems Sustainability and Environmental Policies in Modern Economies* (pp. 15–31). Hershey, PA: IGI Global. doi:10.4018/978-1-5225-3631-4.ch002

Gonzalez-Feliu, J. (2018). Sustainability Evaluation of Green Urban Logistics Systems: Literature Overview and Proposed Framework. In A. Paul, D. Bhattacharyya, & S. Anand (Eds.), *Green Initiatives for Business Sustainability and Value Creation* (pp. 103–134). Hershey, PA: IGI Global. doi:10.4018/978-1-5225-2662-9.ch005

Goodland, R. (2017). A Fresh Look at Livestock Greenhouse Gas Emissions and Mitigation. In I. Management Association (Ed.), Natural Resources Management: Concepts, Methodologies, Tools, and Applications (pp. 124-139). Hershey, PA: IGI Global. doi:10.4018/978-1-5225-0803-8.ch006

Goundar, S., & Appana, S. (2018). Mainstreaming Development Policies for Climate Change in Fiji: A Policy Gap Analysis and the Role of ICTs. In I. Management Association (Ed.), Sustainable Development: Concepts, Methodologies, Tools, and Applications (pp. 402-432). Hershey, PA: IGI Global. doi:10.4018/978-1-5225-3817-2.ch020

Granell-Canut, C., & Aguilar-Moreno, E. (2018). Geospatial Influence in Science Mapping. In M. Khosrow-Pour, D.B.A. (Ed.), Encyclopedia of Information Science and Technology, Fourth Edition (pp. 3473-3483). Hershey, PA: IGI Global. doi:10.4018/978-1-5225-2255-3.ch302

Grigelis, A., Blažauskas, N., Gelumbauskaitė, L. Ž., Gulbinskas, S., Suzdalev, S., & Ferrarin, C. (2017). Marine Environment Data Management Related to the Human Activity in the South-Eastern Baltic Sea (The Lithuanian Sector). In P. Diviacco, A. Leadbetter, & H. Glaves (Eds.), *Oceanographic and Marine Cross-Domain Data Management for Sustainable Development* (pp. 282–302). Hershey, PA: IGI Global. doi:10.4018/978-1-5225-0700-0.ch012

Guma, I. P., Rwashana, A. S., & Oyo, B. (2018). Food Security Policy Analysis Using System Dynamics: The Case of Uganda. *International Journal of Information Technologies and Systems Approach*, *11*(1), 72–90. doi:10.4018/IJITSA.2018010104

Guma, I. P., Rwashana, A. S., & Oyo, B. (2018). Food Security Indicators for Subsistence Farmers Sustainability: A System Dynamics Approach. *International Journal of System Dynamics Applications*, *7*(1), 45–64. doi:10.4018/IJSDA.2018010103

Gupta, P., & Goyal, S. (2017). Wildlife Habitat Evaluation. In A. Santra & S. Mitra (Eds.), *Remote Sensing Techniques and GIS Applications in Earth and Environmental Studies* (pp. 258–264). Hershey, PA: IGI Global. doi:10.4018/978-1-5225-1814-3.ch013

Hanson, T., & Hildebrand, E. (2018). GPS Travel Diaries in Rural Transportation Research: A Focus on Older Drivers. In I. Management Association (Ed.), Intelligent Transportation and Planning: Breakthroughs in Research and Practice (pp. 609-625). Hershey, PA: IGI Global. doi:10.4018/978-1-5225-5210-9.ch027

Hartman, M. B. (2017). Research-Based Climate Change Public Education Programs. In I. Management Association (Ed.), Natural Resources Management: Concepts, Methodologies, Tools, and Applications (pp. 992-1003). Hershey, PA: IGI Global. doi:10.4018/978-1-5225-0803-8.ch046

Hashim, N. (2017). Zanzibari Seaweed: Global Climate Change and the Promise of Adaptation. In I. Management Association (Ed.), Natural Resources Management: Concepts, Methodologies, Tools, and Applications (pp. 365-391). Hershey, PA: IGI Global. doi:10.4018/978-1-5225-0803-8.ch019

Herrera, J. E., Argüello, L. V., Gonzalez-Feliu, J., & Jaimes, W. A. (2017). Decision Support System Design Requirements, Information Management, and Urban Logistics Efficiency: Case Study of Bogotá, Colombia. In G. Jamil, A. Soares, & C. Pessoa (Eds.), *Handbook of Research on Information Management for Effective Logistics and Supply Chains* (pp. 223–238). Hershey, PA: IGI Global. doi:10.4018/978-1-5225-0973-8.ch012

Huizinga, T., Ayanso, A., Smoor, M., & Wronski, T. (2017). Exploring Insurance and Natural Disaster Tweets Using Text Analytics. *International Journal of Business Analytics*, *4*(1), 1–17. doi:10.4018/IJBAN.2017010101

Hung, K., Kalantari, M., & Rajabifard, A. (2017). An Integrated Method for Assessing the Text Content Quality of Volunteered Geographic Information in Disaster Management. *International Journal of Information Systems for Crisis Response and Management*, *9*(2), 1–17. doi:10.4018/IJISCRAM.2017040101

Husnain, A., & Avdic, A. (2018). Identifying the Contemporary Status of E-Service Sustainability Research. In I. Management Association (Ed.), Sustainable Development: Concepts, Methodologies, Tools, and Applications (pp. 467-485). Hershey, PA: IGI Global. doi:10.4018/978-1-5225-3817-2.ch022

Iarossi, M. P., & Ferro, L. (2017). "The Past is Never Dead. It's Not Even Past": Virtual Archaeological Promenade. In A. Ippolito & M. Cigola (Eds.), *Handbook of Research on Emerging Technologies for Digital Preservation and Information Modeling* (pp. 228–255). Hershey, PA: IGI Global. doi:10.4018/978-1-5225-0680-5.ch010

Ignjatijević, S., & Cvijanović, D. (2018). Analysis of Serbian Honey Production and Exports. In *Exploring the Global Competitiveness of Agri-Food Sectors and Serbia's Dominant Presence: Emerging Research and Opportunities* (pp. 109–139). Hershey, PA: IGI Global. doi:10.4018/978-1-5225-2762-6.ch005

Jana, S. K., & Karmakar, A. K. (2017). Globalization, Governance, and Food Security: The Case of BRICS. In I. Management Association (Ed.), Natural Resources Management: Concepts, Methodologies, Tools, and Applications (pp. 692-712). Hershey, PA: IGI Global. doi:10.4018/978-1-5225-0803-8.ch032

Jana, S. K., & Karmakar, A. K. (2017). Food Security in Asia: Is There Convergence? In I. Management Association (Ed.), Natural Resources Management: Concepts, Methodologies, Tools, and Applications (pp. 109-123). Hershey, PA: IGI Global. doi:10.4018/978-1-5225-0803-8.ch005

John, J., & Kumar, S. (2018). A Locational Decision Making Framework for Shipbreaking Under Multiple Criteria. In I. Management Association (Ed.), Operations and Service Management: Concepts, Methodologies, Tools, and Applications (pp. 504-527). Hershey, PA: IGI Global. doi:10.4018/978-1-5225-3909-4.ch024

John, J., & Srivastava, R. K. (2018). Decision Insights for Shipbreaking using Environmental Impact Assessment: Review and Perspectives. *International Journal of Strategic Decision Sciences*, *9*(1), 45–62. doi:10.4018/IJSDS.2018010104

Joshi, Y., & Rahman, Z. (2018). Determinants of Sustainable Consumption Behaviour: Review and Conceptual Framework. In A. Paul, D. Bhattacharyya, & S. Anand (Eds.), *Green Initiatives for Business Sustainability and Value Creation* (pp. 239–262). Hershey, PA: IGI Global. doi:10.4018/978-1-5225-2662-9.ch011

Juma, D. W., Reuben, M., Wang, H., & Li, F. (2018). Adaptive Coevolution: Realigning the Water Governance Regime to the Changing Climate. In I. Management Association (Ed.), Hydrology and Water Resource Management: Breakthroughs in Research and Practice (pp. 346-357). Hershey, PA: IGI Global. doi:10.4018/978-1-5225-3427-3.ch014

K., S., & Tripathy, B. K. (2018). Neighborhood Rough-Sets-Based Spatial Data Analytics. In M. Khosrow-Pour, D.B.A. (Ed.), *Encyclopedia of Information Science and Technology, Fourth Edition* (pp. 1835-1844). Hershey, PA: IGI Global. doi:10.4018/978-1-5225-2255-3.ch160

Kabir, F. (2018). Towards a More Gender-Inclusive Climate Change Policy. In N. Mahtab, T. Haque, I. Khan, M. Islam, & I. Wahid (Eds.), *Handbook of Research on Women's Issues and Rights in the Developing World* (pp. 354–369). Hershey, PA: IGI Global. doi:10.4018/978-1-5225-3018-3.ch022

Kabir, F. (2018). Towards a More Gender-Inclusive Climate Change Policy. In I. Management Association (Ed.), Climate Change and Environmental Concerns: Breakthroughs in Research and Practice (pp. 525-540). Hershey, PA: IGI Global. doi:10.4018/978-1-5225-5487-5.ch027

Kanyamuka, J. S., Jumbe, C. B., & Ricker-Gilbert, J. (2018). Making Agricultural Input Subsidies More Effective and Profitable in Africa: The Role of Complementary Interventions. In A. Obayelu (Ed.), *Food Systems Sustainability and Environmental Policies in Modern Economies* (pp. 172–187). Hershey, PA: IGI Global. doi:10.4018/978-1-5225-3631-4.ch008

Karimi, H., & Gholamrezafahimi, F. (2017). Study of Integrated Coastal Zone Management and Its Environmental Effects: A Case of Iran. In R. Singh, A. Singh, & V. Srivastava (Eds.), *Environmental Issues Surrounding Human Overpopulation* (pp. 64–88). Hershey, PA: IGI Global. doi:10.4018/978-1-5225-1683-5.ch004

Kaya, I. R., Hutabarat, J., & Bambang, A. N. (2018). "Sasi": A New Path to Sustain Seaweed Farming From Up-Stream to Down-Stream in Kotania Bay, Molucass. *International Journal of Social Ecology and Sustainable Development*, *9*(2), 28–36. doi:10.4018/IJSESD.2018040103

Khader, V. (2018). Technologies for Food, Health, Livelihood, and Nutrition Security. In I. Management Association (Ed.), Food Science and Nutrition: Breakthroughs in Research and Practice (pp. 94-112). Hershey, PA: IGI Global. doi:10.4018/978-1-5225-5207-9.ch005

Kocadağlı, A. Y. (2017). The Temporal and Spatial Development of Organic Agriculture in Turkey. In W. Ganpat, R. Dyer, & W. Isaac (Eds.), *Agricultural Development and Food Security in Developing Nations* (pp. 130–156). Hershey, PA: IGI Global. doi:10.4018/978-1-5225-0942-4.ch006

Koundouri, P., Giannouli, A., & Souliotis, I. (2017). An Integrated Approach for Sustainable Environmental and Socio-Economic Development Using Offshore Infrastructure. In I. Management Association (Ed.), Renewable and Alternative Energy: Concepts, Methodologies, Tools, and Applications (pp. 1581-1601). Hershey, PA: IGI Global. doi:10.4018/978-1-5225-1671-2.ch056

Kumar, A., & Dash, M. K. (2017). Sustainability and Future Generation Infrastructure on Digital Platform: A Study of Generation Y. In N. Ray (Ed.), *Business Infrastructure for Sustainability in Developing Economies* (pp. 124–142). Hershey, PA: IGI Global. doi:10.4018/978-1-5225-2041-2.ch007

Kumar, A., Mukherjee, A. B., & Krishna, A. P. (2017). Application of Conventional Data Mining Techniques and Web Mining to Aid Disaster Management. In A. Kumar (Ed.), *Web Usage Mining Techniques and Applications Across Industries* (pp. 138–167). Hershey, PA: IGI Global. doi:10.4018/978-1-5225-0613-3.ch006

Kumar, C. P. (2017). Impact of Climate Change on Groundwater Resources. In I. Management Association (Ed.), Natural Resources Management: Concepts, Methodologies, Tools, and Applications (pp. 1094-1120). Hershey, PA: IGI Global. doi:10.4018/978-1-5225-0803-8.ch052

Kumari, S., & Patil, Y. (2017). Achieving Climate Smart Agriculture with a Sustainable Use of Water: A Conceptual Framework for Sustaining the Use of Water for Agriculture in the Era of Climate Change. In P. Rao & Y. Patil (Eds.), *Reconsidering the Impact of Climate Change on Global Water Supply, Use, and Management* (pp. 122–143). Hershey, PA: IGI Global. doi:10.4018/978-1-5225-1046-8.ch008

Kumari, S., & Patil, Y. (2018). Achieving Climate Smart Agriculture With a Sustainable Use of Water: A Conceptual Framework for Sustaining the Use of Water for Agriculture in the Era of Climate Change. In I. Management Association (Ed.), Climate Change and Environmental Concerns: Breakthroughs in Research and Practice (pp. 111-133). Hershey, PA: IGI Global. doi:10.4018/978-1-5225-5487-5.ch006

Kursah, M. B. (2017). Least-Cost Pipeline using Geographic Information System: The Limit to Technicalities. *International Journal of Applied Geospatial Research, 8*(3), 1–15. doi:10.4018/ijagr.2017070101

Lahiri, S., Ghosh, D., & Bhakta, J. N. (2017). Role of Microbes in Eco-Remediation of Perturbed Aquatic Ecosystem. In J. Bhakta (Ed.), *Handbook of Research on Inventive Bioremediation Techniques* (pp. 70–107). Hershey, PA: IGI Global. doi:10.4018/978-1-5225-2325-3.ch004

Lallo, C. H., Smalling, S., Facey, A., & Hughes, M. (2018). The Impact of Climate Change on Small Ruminant Performance in Caribbean Communities. In I. Management Association (Ed.), Climate Change and Environmental Concerns: Breakthroughs in Research and Practice (pp. 193-218). Hershey, PA: IGI Global. doi:10.4018/978-1-5225-5487-5.ch010

Laurini, R. (2017). Nature of Geographic Knowledge Bases. In S. Faiz & K. Mahmoudi (Eds.), *Handbook of Research on Geographic Information Systems Applications and Advancements* (pp. 29–60). Hershey, PA: IGI Global. doi:10.4018/978-1-5225-0937-0.ch002

Lawrence, J., Simpson, L., & Piggott, A. (2017). Protected Agriculture: A Climate Change Adaptation for Food and Nutrition Security. In I. Management Association (Ed.), Natural Resources Management: Concepts, Methodologies, Tools, and Applications (pp. 140-158). Hershey, PA: IGI Global. doi:10.4018/978-1-5225-0803-8.ch007

Leadbetter, A., Cheatham, M., Shepherd, A., & Thomas, R. (2017). Linked Ocean Data 2.0. In P. Diviacco, A. Leadbetter, & H. Glaves (Eds.), *Oceanographic and Marine Cross-Domain Data Management for Sustainable Development* (pp. 69–99). Hershey, PA: IGI Global. doi:10.4018/978-1-5225-0700-0.ch004

Leipnik, M. R., & Mehta, S. S. (2005). Geographic Information Systems (GIS) in E-Marketing. In I. Clarke III & T. Flaherty (Eds.), *Advances in Electronic Marketing* (pp. 193–210). Hershey, PA: IGI Global. doi:10.4018/978-1-59140-321-0.ch011

Lucas, M. R., Rego, C., Vieira, C., & Vieira, I. (2017). Proximity and Cooperation for Innovative Regional Development: The Case of the Science and Technology Park of Alentejo. In L. Carvalho (Ed.), *Handbook of Research on Entrepreneurial Development and Innovation Within Smart Cities* (pp. 199–228). Hershey, PA: IGI Global. doi:10.4018/978-1-5225-1978-2.ch010

Ma, X., Beaulieu, S. E., Fu, L., Fox, P., Di Stefano, M., & West, P. (2017). Documenting Provenance for Reproducible Marine Ecosystem Assessment in Open Science. In P. Diviacco, A. Leadbetter, & H. Glaves (Eds.), *Oceanographic and Marine Cross-Domain Data Management for Sustainable Development* (pp. 100–126). Hershey, PA: IGI Global. doi:10.4018/978-1-5225-0700-0.ch005

Mabe, L. K., & Oladele, O. I. (2017). Application of Information Communication Technologies for Agricultural Development through Extension Services: A Review. In T. Tossy (Ed.), *Information Technology Integration for Socio-Economic Development* (pp. 52–101). Hershey, PA: IGI Global. doi:10.4018/978-1-5225-0539-6.ch003

Malomo, B. I. (2018). A Review of Psychological Resilience as a Response to Natural Hazards in Nigeria. In A. Eneanya (Ed.), *Handbook of Research on Environmental Policies for Emergency Management and Public Safety* (pp. 147–165). Hershey, PA: IGI Global. doi:10.4018/978-1-5225-3194-4.ch008

Manchiraju, S. (2018). Predicting Behavioral Intentions Toward Sustainable Fashion Consumption: A Comparison of Attitude-Behavior and Value-Behavior Consistency Models. In I. Management Association (Ed.), Fashion and Textiles: Breakthroughs in Research and Practice (pp. 1-21). Hershey, PA: IGI Global. doi:10.4018/978-1-5225-3432-7.ch001

Manzella, G. M., Bartolini, R., Bustaffa, F., D'Angelo, P., De Mattei, M., Frontini, F., ... Spada, A. (2017). Semantic Search Engine for Data Management and Sustainable Development: Marine Planning Service Platform. In P. Diviacco, A. Leadbetter, & H. Glaves (Eds.), *Oceanographic and Marine Cross-Domain Data Management for Sustainable Development* (pp. 127–154). Hershey, PA: IGI Global. doi:10.4018/978-1-5225-0700-0.ch006

Mbonigaba, J. (2018). Comparing the Effects of Unsustainable Production and Consumption of Food on Health and Policy Across Developed and Less Developed Countries. In A. Obayelu (Ed.), *Food Systems Sustainability and Environmental Policies in Modern Economies* (pp. 124–158). Hershey, PA: IGI Global. doi:10.4018/978-1-5225-3631-4.ch006

McKeown, A. E. (2017). Nurses, Healthcare, and Environmental Pollution and Solutions: Breaking the Cycle of Harm. In I. Management Association (Ed.), Natural Resources Management: Concepts, Methodologies, Tools, and Applications (pp. 392-415). Hershey, PA: IGI Global. doi:10.4018/978-1-5225-0803-8.ch020

Mennecke, B. E., & West, L. A. Jr. (2001). Geographic Information Systems in Developing Countries: Issues in Data Collection, Implementation and Management. *Journal of Global Information Management*, *9*(4), 44–54. doi:10.4018/jgim.2001100103

Mili, B., Barua, A., & Katyaini, S. (2017). Climate Change and Adaptation through the Lens of Capability Approach: A Case Study from Darjeeling, Eastern Himalaya. In I. Management Association (Ed.), Natural Resources Management: Concepts, Methodologies, Tools, and Applications (pp. 1351-1365). Hershey, PA: IGI Global. doi:10.4018/978-1-5225-0803-8.ch064

Mir, S. A., Shah, M. A., Mir, M. M., & Iqbal, U. (2017). New Horizons of Nanotechnology in Agriculture and Food Processing Industry. In B. Nayak, A. Nanda, & M. Bhat (Eds.), *Integrating Biologically-Inspired Nanotechnology into Medical Practice* (pp. 230–258). Hershey, PA: IGI Global. doi:10.4018/978-1-5225-0610-2.ch009

Mujere, N., & Moyce, W. (2017). Climate Change Impacts on Surface Water Quality. In W. Ganpat & W. Isaac (Eds.), *Environmental Sustainability and Climate Change Adaptation Strategies* (pp. 322–340). Hershey, PA: IGI Global. doi:10.4018/978-1-5225-1607-1.ch012

Mukherjee, A. B., Krishna, A. P., & Patel, N. (2018). Geospatial Technology for Urban Sciences. In *Geospatial Technologies in Urban System Development: Emerging Research and Opportunities* (pp. 99–120). Hershey, PA: IGI Global. doi:10.4018/978-1-5225-3683-3.ch005

Nagarajan, S. K., & Sangaiah, A. K. (2017). Vegetation Index: Ideas, Methods, Influences, and Trends. In N. Kumar, A. Sangaiah, M. Arun, & S. Anand (Eds.), *Advanced Image Processing Techniques and Applications* (pp. 347–386). Hershey, PA: IGI Global. doi:10.4018/978-1-5225-2053-5.ch016

Naraine, L., & Meehan, K. (2017). Strengthening Food Security with Sustainable Practices by Smallholder Farmers in Lesser Developed Economies. In W. Ganpat, R. Dyer, & W. Isaac (Eds.), *Agricultural Development and Food Security in Developing Nations* (pp. 57–81). Hershey, PA: IGI Global. doi:10.4018/978-1-5225-0942-4.ch003

Naraine, L., & Meehan, K. (2017). Strengthening Food Security with Sustainable Practices by Smallholder Farmers in Lesser Developed Economies. In W. Ganpat, R. Dyer, & W. Isaac (Eds.), *Agricultural Development and Food Security in Developing Nations* (pp. 57–81). Hershey, PA: IGI Global. doi:10.4018/978-1-5225-0942-4.ch003

Naraine, L., & Meehan, K. (2017). Strengthening Food Security with Sustainable Practices by Smallholder Farmers in Lesser Developed Economies. In W. Ganpat, R. Dyer, & W. Isaac (Eds.), *Agricultural Development and Food Security in Developing Nations* (pp. 57–81). Hershey, PA: IGI Global. doi:10.4018/978-1-5225-0942-4.ch003

Neufeld, D. J., & Griffith, S. (2006). Isobord's Geographic Information System (GIS) Solution. In M. Khosrow-Pour, D.B.A. (Ed.), Cases on Database Technologies and Applications (pp. 64-80). Hershey, PA: IGI Global. doi:10.4018/978-1-59904-399-9.ch004

Nikolaou, K., Tsakiridou, E., Anastasiadis, F., & Mattas, K. (2017). Exploring Alternative Distribution Channels of Agricultural Products. *International Journal of Food and Beverage Manufacturing and Business Models, 2*(2), 36–66. doi:10.4018/IJFBMBM.2017070103

Nishat, K. J., & Rahman, M. S. (2018). Disaster, Vulnerability, and Violence Against Women: Global Findings and a Research Agenda for Bangladesh. In N. Mahtab, T. Haque, I. Khan, M. Islam, & I. Wahid (Eds.), *Handbook of Research on Women's Issues and Rights in the Developing World* (pp. 235–250). Hershey, PA: IGI Global. doi:10.4018/978-1-5225-3018-3.ch014

O'Hara, S., Jones, D., & Trobman, H. B. (2018). Building an Urban Food System Through UDC Food Hubs. In A. Burtin, J. Fleming, & P. Hampton-Garland (Eds.), *Changing Urban Landscapes Through Public Higher Education* (pp. 116–143). Hershey, PA: IGI Global. doi:10.4018/978-1-5225-3454-9.ch006

Obayelu, A. E. (2018). Integrating Environment, Food Systems, and Sustainability in Feeding the Growing Population in Developing Countries. In A. Obayelu (Ed.), *Food Systems Sustainability and Environmental Policies in Modern Economies* (pp. 1–14). Hershey, PA: IGI Global. doi:10.4018/978-1-5225-3631-4.ch001

Othman, R., Nath, N., & Laswad, F. (2018). Environmental Reporting and Accounting: Sustainability Hybridisation. In G. Azevedo, J. da Silva Oliveira, R. Marques, & A. Ferreira (Eds.), *Handbook of Research on Modernization and Accountability in Public Sector Management* (pp. 130–158). Hershey, PA: IGI Global. doi:10.4018/978-1-5225-3731-1.ch007

Padigala, B. S. (2018). Traditional Water Management System for Climate Change Adaptation in Mountain Ecosystems. In I. Management Association (Ed.), Climate Change and Environmental Concerns: Breakthroughs in Research and Practice (pp. 630-655). Hershey, PA: IGI Global. doi:10.4018/978-1-5225-5487-5.ch033

Panda, C. K. (2018). Mobile Phone Usage in Agricultural Extension in India: The Current and Future Perspective. In F. Mtenzi, G. Oreku, D. Lupiana, & J. Yonazi (Eds.), *Mobile Technologies and Socio-Economic Development in Emerging Nations* (pp. 1–21). Hershey, PA: IGI Global. doi:10.4018/978-1-5225-4029-8.ch001

Pandian, S. L., Yarrakula, K., & Chaudhury, P. (2018). GIS-Based Decision Support System for Village Level: A Case Study in Andhra Pradesh. In S. Chaudhuri & N. Ray (Eds.), *GIS Applications in the Tourism and Hospitality Industry* (pp. 275–295). Hershey, PA: IGI Global. doi:10.4018/978-1-5225-5088-4.ch012

Quaranta, G., & Salvia, R. (2017). Social-Based Product Innovation and Governance in The Milk Sector: The Case of Carciocacio and Innonatura. In T. Tarnanidis, M. Vlachopoulou, & J. Papathanasiou (Eds.), *Driving Agribusiness With Technology Innovations* (pp. 293–310). Hershey, PA: IGI Global. doi:10.4018/978-1-5225-2107-5.ch015

Rahman, M. K., Schmidlin, T. W., Munro-Stasiuk, M. J., & Curtis, A. (2017). Geospatial Analysis of Land Loss, Land Cover Change, and Landuse Patterns of Kutubdia Island, Bangladesh. *International Journal of Applied Geospatial Research*, *8*(2), 45–60. doi:10.4018/IJAGR.2017040104

Rajack-Talley, T. A. (2017). Agriculture, Trade Liberalization and Poverty in the ACP Countries. In W. Ganpat, R. Dyer, & W. Isaac (Eds.), *Agricultural Development and Food Security in Developing Nations* (pp. 1–20). Hershey, PA: IGI Global. doi:10.4018/978-1-5225-0942-4.ch001

Rajamanickam, S. (2018). Exploring Landscapes in Regional Convergence: Environment and Sustainable Development in South Asia. In I. Management Association (Ed.), Sustainable Development: Concepts, Methodologies, Tools, and Applications (pp. 1051-1087). Hershey, PA: IGI Global. doi:10.4018/978-1-5225-3817-2.ch047

Rizvi, S. M., & Dearden, A. (2018). KHETI: ICT Solution for Agriculture Extension and Its Replication in Open and Distance Learning. In U. Pandey & V. Indrakanti (Eds.), *Open and Distance Learning Initiatives for Sustainable Development* (pp. 163–174). Hershey, PA: IGI Global. doi:10.4018/978-1-5225-2621-6.ch008

Roşu, L., & Macarov, L. I. (2017). Management of Drought and Floods in the Dobrogea Region. In I. Management Association (Ed.), Agri-Food Supply Chain Management: Breakthroughs in Research and Practice (pp. 372-403). Hershey, PA: IGI Global. doi:10.4018/978-1-5225-1629-3.ch016

Rouzbehani, K., & Rouzbehani, S. (2018). Mapping Women's World: GIS and the Case of Breast Cancer in the US. *International Journal of Public Health Management and Ethics*, *3*(1), 14–25. doi:10.4018/IJPHME.2018010102

Roy, D. (2018). Success Factors of Adoption of Mobile Applications in Rural India: Effect of Service Characteristics on Conceptual Model. In M. Khosrow-Pour, D.B.A. (Ed.), Green Computing Strategies for Competitive Advantage and Business Sustainability (pp. 211-238). Hershey, PA: IGI Global. doi:10.4018/978-1-5225-5017-4.ch010

Sajeva, M., Lemon, M., & Sahota, P. S. (2017). Governance for Food Security: A Framework for Social Learning and Scenario Building. *International Journal of Food and Beverage Manufacturing and Business Models*, *2*(2), 67–84. doi:10.4018/IJFBMBM.2017070104

Sallam, G. A., Youssef, T., Embaby, M. E., & Shaltot, F. (2011). Using Geographic Information System to Infollow the Fertilizers Pollution Migration. In I. Management Association (Ed.), Green Technologies: Concepts, Methodologies, Tools and Applications (pp. 564-586). Hershey, PA: IGI Global. doi:10.4018/978-1-60960-472-1.ch312

Sambhanthan, A., & Potdar, V. (2017). A Study of the Parameters Impacting Sustainability in Information Technology Organizations. *International Journal of Knowledge-Based Organizations*, *7*(3), 27–39. doi:10.4018/IJKBO.2017070103

Sanga, C., Kalungwizi, V. J., & Msuya, C. P. (2017). Bridging Gender Gaps in Provision of Agricultural Extension Service Using ICT: Experiences from Sokoine University of Agriculture (SUA) Farmer Voice Radio (FVR) Project in Tanzania. In I. Management Association (Ed.), Discrimination and Diversity: Concepts, Methodologies, Tools, and Applications (pp. 682-697). Hershey, PA: IGI Global. doi:10.4018/978-1-5225-1933-1.ch031

Santra, A., & Mitra, D. (2017). Role of Remote Sensing in Potential Fishing Zone Forecast. In A. Santra & S. Mitra (Eds.), *Remote Sensing Techniques and GIS Applications in Earth and Environmental Studies* (pp. 243–257). Hershey, PA: IGI Global. doi:10.4018/978-1-5225-1814-3.ch012

Schaap, D. (2017). SeaDataNet: Towards a Pan-European Infrastructure for Marine and Ocean Data Management. In P. Diviacco, A. Leadbetter, & H. Glaves (Eds.), *Oceanographic and Marine Cross-Domain Data Management for Sustainable Development* (pp. 155–177). Hershey, PA: IGI Global. doi:10.4018/978-1-5225-0700-0.ch007

Seckin-Celik, T. (2017). Sustainability Reporting and Sustainability in the Turkish Business Context. In U. Akkucuk (Ed.), *Ethics and Sustainability in Global Supply Chain Management* (pp. 115–132). Hershey, PA: IGI Global. doi:10.4018/978-1-5225-2036-8.ch006

Segbefia, A. Y., Barnes, V. R., Akpalu, L. A., & Mensah, M. (2018). Environmental Location Assessment for Seaweed Cultivation in Ghana: A Spatial Multi-Criteria Approach. *International Journal of Applied Geospatial Research*, *9*(1), 51–64. doi:10.4018/IJAGR.2018010104

Sen, Y. (2018). How to Manage Sustainability: A Framework for Corporate Sustainability Tools. In I. Management Association (Ed.), Sustainable Development: Concepts, Methodologies, Tools, and Applications (pp. 568-589). Hershey, PA: IGI Global. doi:10.4018/978-1-5225-3817-2.ch026

Shamshiry, E., Abdulai, A. M., Mokhtar, M. B., & Komoo, I. (2015). Regional Landfill Site Selection with GIS and Analytical Hierarchy Process Techniques: A Case Study of Langkawi Island, Malaysia. In P. Thomas, M. Srihari, & S. Kaur (Eds.), *Handbook of Research on Cultural and Economic Impacts of the Information Society* (pp. 248–282). Hershey, PA: IGI Global. doi:10.4018/978-1-4666-8598-7.ch011

Sharma, Y. K., Mangla, S. K., Patil, P. P., & Uniyal, S. (2018). Analyzing Sustainable Food Supply Chain Management Challenges in India. In M. Ram & J. Davim (Eds.), *Soft Computing Techniques and Applications in Mechanical Engineering* (pp. 162–180). Hershey, PA: IGI Global. doi:10.4018/978-1-5225-3035-0.ch008

Silvestrelli, P. (2018). The Impact of Events: To Which Extent Are They Sustainable for Tourist Destinations? Some Evidences From Expo Milano 2015. In M. Risso & S. Testarmata (Eds.), *Value Sharing for Sustainable and Inclusive Development* (pp. 185–204). Hershey, PA: IGI Global. doi:10.4018/978-1-5225-3147-0.ch008

Silvius, G. (2018). Sustainability Evaluation of IT/IS Projects. In I. Management Association (Ed.), Sustainable Development: Concepts, Methodologies, Tools, and Applications (pp. 26-40). Hershey, PA: IGI Global. doi:10.4018/978-1-5225-3817-2.ch002

Singh, R., Srivastava, P., Singh, P., Upadhyay, S., & Raghubanshi, A. S. (2017). Human Overpopulation and Food Security: Challenges for the Agriculture Sustainability. In R. Singh, A. Singh, & V. Srivastava (Eds.), *Environmental Issues Surrounding Human Overpopulation* (pp. 12–39). Hershey, PA: IGI Global. doi:10.4018/978-1-5225-1683-5.ch002

Singh, R., Srivastava, P., Singh, P., Upadhyay, S., & Raghubanshi, A. S. (2017). Human Overpopulation and Food Security: Challenges for the Agriculture Sustainability. In R. Singh, A. Singh, & V. Srivastava (Eds.), *Environmental Issues Surrounding Human Overpopulation* (pp. 12–39). Hershey, PA: IGI Global. doi:10.4018/978-1-5225-1683-5.ch002

Srivastava, N. (2017). Climate Change Mitigation: Collective Efforts and Responsibly. In I. Management Association (Ed.), Natural Resources Management: Concepts, Methodologies, Tools, and Applications (pp. 64-76). Hershey, PA: IGI Global. doi:10.4018/978-1-5225-0803-8.ch003

Stanganelli, M., & Gerundo, C. (2017). Understanding the Role of Urban Morphology and Green Areas Configuration During Heat Waves. *International Journal of Agricultural and Environmental Information Systems*, *8*(2), 50–64. doi:10.4018/IJAEIS.2017040104

Stewart, M. K., Hagood, D., & Ching, C. C. (2017). Virtual Games and Real-World Communities: Environments that Constrain and Enable Physical Activity in Games for Health. *International Journal of Game-Based Learning*, *7*(1), 1–19. doi:10.4018/IJGBL.2017010101

Stone, R. J. (2017). Modelling the Frequency of Tropical Cyclones in the Lower Caribbean Region. In W. Ganpat & W. Isaac (Eds.), *Environmental Sustainability and Climate Change Adaptation Strategies* (pp. 341–349). Hershey, PA: IGI Global. doi:10.4018/978-1-5225-1607-1.ch013

Syed, A., & Jabeen, U. A. (2018). Climate Change Impact on Agriculture and Food Security. In A. Eneanya (Ed.), *Handbook of Research on Environmental Policies for Emergency Management and Public Safety* (pp. 223–237). Hershey, PA: IGI Global. doi:10.4018/978-1-5225-3194-4.ch012

Tam, G. C. (2017). The Global View of Sustainability. In *Managerial Strategies and Green Solutions for Project Sustainability* (pp. 1–24). Hershey, PA: IGI Global. doi:10.4018/978-1-5225-2371-0.ch001

Tam, G. C. (2017). Understanding Project Sustainability. In *Managerial Strategies and Green Solutions for Project Sustainability* (pp. 110–139). Hershey, PA: IGI Global. doi:10.4018/978-1-5225-2371-0.ch005

Tam, G. C. (2017). Perspectives on Sustainability. In *Managerial Strategies and Green Solutions for Project Sustainability* (pp. 53–76). Hershey, PA: IGI Global. doi:10.4018/978-1-5225-2371-0.ch003

Tang, M., & Karunanithi, A. T. (2018). Visual Logic Maps (vLms). In *Advanced Concept Maps in STEM Education: Emerging Research and Opportunities* (pp. 108–149). Hershey, PA: IGI Global. doi:10.4018/978-1-5225-2184-6.ch005

Taşçıoğlu, M., & Yener, D. (2018). The Value and Scope of GIS in Marketing and Tourism Management. In S. Chaudhuri & N. Ray (Eds.), *GIS Applications in the Tourism and Hospitality Industry* (pp. 189–211). Hershey, PA: IGI Global. doi:10.4018/978-1-5225-5088-4.ch009

Tianming, G. (2018). Food Security and Rural Development on Emerging Markets of Northeast Asia: Cases of Chinese North and Russian Far East. In V. Erokhin (Ed.), *Establishing Food Security and Alternatives to International Trade in Emerging Economies* (pp. 155–176). Hershey, PA: IGI Global. doi:10.4018/978-1-5225-2733-6.ch008

Tiftikçigil, B. Y., Yaşgül, Y. S., & Güriş, B. (2017). Sustainability of Foreign Trade Deficit in Energy: The Case of Turkey. In N. Ray (Ed.), *Business Infrastructure for Sustainability in Developing Economies* (pp. 94–109). Hershey, PA: IGI Global. doi:10.4018/978-1-5225-2041-2.ch005

Tiwari, S., Vaish, B., & Singh, P. (2018). Population and Global Food Security: Issues Related to Climate Change. In I. Management Association (Ed.), Climate Change and Environmental Concerns: Breakthroughs in Research and Practice (pp. 41-64). Hershey, PA: IGI Global. doi:10.4018/978-1-5225-5487-5.ch003

Toujani, A., & Achour, H. (2018). A Data Mining Framework for Forest Fire Mapping. In I. Management Association (Ed.), Information Retrieval and Management: Concepts, Methodologies, Tools, and Applications (pp. 771-794). Hershey, PA: IGI Global. doi:10.4018/978-1-5225-5191-1.ch033

Tripathy, B., & K., S. B. (2017). Rough Fuzzy Set Theory and Neighbourhood Approximation Based Modelling for Spatial Epidemiology. In I. Management Association (Ed.), *Public Health and Welfare: Concepts, Methodologies, Tools, and Applications* (pp. 1257-1268). Hershey, PA: IGI Global. doi:10.4018/978-1-5225-1674-3.ch058

Trukhachev, A. (2017). New Approaches to Regional Branding through Green Production and Utilization of Existing Natural Advantages. In I. Management Association (Ed.), Advertising and Branding: Concepts, Methodologies, Tools, and Applications (pp. 1758-1778). Hershey, PA: IGI Global. doi:10.4018/978-1-5225-1793-1.ch081

Tsobanoglou, G. O., & Vlachopoulou, E. I. (2017). Social-Ecological Systems in Local Fisheries Communities. In G. Korres, E. Kourliouros, & M. Michailidis (Eds.), *Handbook of Research on Policies and Practices for Sustainable Economic Growth and Regional Development* (pp. 306–316). Hershey, PA: IGI Global. doi:10.4018/978-1-5225-2458-8.ch026

Tsou, M. (2010). Geographic Information Retrieval and Text Mining on Chinese Tourism Web Pages. *International Journal of Information Technology and Web Engineering*, *5*(1), 56–75. doi:10.4018/jitwe.2010010104

Tuydes-Yaman, H., & Karatas, P. (2018). Evaluation of Walkability and Pedestrian Level of Service. In I. Management Association (Ed.), Intelligent Transportation and Planning: Breakthroughs in Research and Practice (pp. 264-291). Hershey, PA: IGI Global. doi:10.4018/978-1-5225-5210-9.ch012

Uddin, S., Chakravorty, S., Ray, A., & Sherpa, K. S. (2018). Optimal Location of Sub-Station Using Q-GIS and Multi-Criteria Decision Making Approach. *International Journal of Decision Support System Technology*, *10*(2), 65–79. doi:10.4018/IJDSST.2018040104

Uddin, S., Chakravorty, S., Sherpa, K. S., & Ray, A. (2018). Power Distribution System Planning Using Q-GIS. *International Journal of Energy Optimization and Engineering*, *7*(2), 61–75. doi:10.4018/IJEOE.2018040103

Uzun, F. V. (2018). Natural Resources Management. In A. Eneanya (Ed.), *Handbook of Research on Environmental Policies for Emergency Management and Public Safety* (pp. 1–21). Hershey, PA: IGI Global. doi:10.4018/978-1-5225-3194-4.ch001

V, M., Agrawal, R., Sharma, V., & T.N., K. (2018). Supply Chain Social Sustainability and Manufacturing. In I. Management Association (Ed.), *Technology Adoption and Social Issues: Concepts, Methodologies, Tools, and Applications* (pp. 226-252). Hershey, PA: IGI Global. doi:10.4018/978-1-5225-5201-7.ch011 van der Vliet-Bakker, J. M. (2017). Environmentally Forced Migration and Human Rights. In C. Akrivopoulou (Ed.), *Defending Human Rights and Democracy in the Era of Globalization* (pp. 146–180). Hershey, PA: IGI Global. doi:10.4018/978-1-5225-0723-9.ch007

Vaskov, A. G., Lin, Z. Y., Tyagunov, M. G., Shestopalova, T. A., & Deryugina, G. V. (2018). Design of Renewable Sources GIS for ASEAN Countries. In V. Kharchenko & P. Vasant (Eds.), *Handbook of Research on Renewable Energy and Electric Resources for Sustainable Rural Development* (pp. 1–25). Hershey, PA: IGI Global. doi:10.4018/978-1-5225-3867-7.ch001

Vãzquez, D. G., & Gil, M. T. (2017). Sustainability in Smart Cities: The Case of Vitoria-Gasteiz (Spain) – A Commitment to a New Urban Paradigm. In L. Carvalho (Ed.), *Handbook of Research on Entrepreneurial Development and Innovation Within Smart Cities* (pp. 248–268). Hershey, PA: IGI Global. doi:10.4018/978-1-5225-1978-2.ch012

Wahab, I. N., & Soonthodu, S. (2018). Geographical Information System in Eco-Tourism. In S. Chaudhuri & N. Ray (Eds.), *GIS Applications in the Tourism and Hospitality Industry* (pp. 61–75). Hershey, PA: IGI Global. doi:10.4018/978-1-5225-5088-4.ch003

Weiss-Randall, D. (2018). Cultivating Environmental Justice. In *Utilizing Innovative Technologies to Address the Public Health Impact of Climate Change: Emerging Research and Opportunities* (pp. 110–143). Hershey, PA: IGI Global. doi:10.4018/978-1-5225-3414-3.ch004

Weiss-Randall, D. (2018). Cultivating Resilience. In *Utilizing Innovative Technologies to Address the Public Health Impact of Climate Change: Emerging Research and Opportunities* (pp. 204–235). Hershey, PA: IGI Global. doi:10.4018/978-1-5225-3414-3.ch007

Weiss-Randall, D. (2018). Climate Change Solutions: Where Do We Go From Here? In *Utilizing Innovative Technologies to Address the Public Health Impact of Climate Change: Emerging Research and Opportunities* (pp. 236–268). Hershey, PA: IGI Global. doi:10.4018/978-1-5225-3414-3.ch008

Whyte, K. P., List, M., Stone, J. V., Grooms, D., Gasteyer, S., Thompson, P. B., . . . Bouri, H. (2018). Uberveillance, Standards, and Anticipation: A Case Study on Nanobiosensors in U.S. Cattle. In I. Management Association (Ed.), Biomedical Engineering: Concepts, Methodologies, Tools, and Applications (pp. 577-596). Hershey, PA: IGI Global. doi:10.4018/978-1-5225-3158-6.ch025

Wulff, E. (2017). Data and Operational Oceanography: A Review in Support of Responsible Fisheries and Aquaculture. In P. Diviacco, A. Leadbetter, & H. Glaves (Eds.), *Oceanographic and Marine Cross-Domain Data Management for Sustainable Development* (pp. 303–324). Hershey, PA: IGI Global. doi:10.4018/978-1-5225-0700-0.ch013

Yener, D. (2017). Geographic Information Systems and Its Applications in Marketing Literature. In S. Faiz & K. Mahmoudi (Eds.), *Handbook of Research on Geographic Information Systems Applications and Advancements* (pp. 158–172). Hershey, PA: IGI Global. doi:10.4018/978-1-5225-0937-0.ch006

About the Contributors

Esther Akumbo Nyam (PhD) is a lecturer in the department of Theatre Arts, Plateau State University Bokkos, Nigcria. She holds a PhD in community theatre and development (Kenyatta University Kenya 2016). Her research focuses on conflicts and teaching that revolves around women, children youths in violent conflicts in Jos, community theatre performances and developmental community oriented projects in Nigeria and Kenya. She has published several articles, book chapters on gender violent conflicts, Theatre for development, children's play series and oral performances.

Asiru Hameed Tunde holds a B.Ed. and M.A in English Language from the University of Ibadan, Nigeria and a Ph.D in English and Linguistics from Kenyatta University, Nairobi. Kenya. He teaches linguistic related courses in the Department of English and French, Umaru Musa Yar'adua University, Katsina, Nigeria. His research interests include: Critical Discourse Analysis, Pragmatics, Semantics, Stylistics and Media Discourse. He has published articles in peer-reviewed international journals; co-edited a book on *Political Discourse in Emergent, Fragile and Failed Democracies* published by IGI Global. He is currently working on another co-edited book on WHEN "HOME" MEANS MORE THAN ONE COUNTRY: *The Discursive (Re) Construction of Identities in Trans-National Migrant Communities.* Hameed has also participated in international conferences such as WACRC, Porto-Novo, 2012, AUYL, Nairobi, 2015, LLS, Bangkok, 2016 and ICHDA Ibadan, 2017.

* * *

Andrew Danjuma Dewan obtained his degrees in Universities in Edinburgh and Salford, both in the United Kingdom. He is currently rounding off his PhD funded by Tertiary Education Trust Fund (TETFUND) Nigeria. He is an experienced lecturer and a former Coordinator/HOD Mass Communication Department, Plateau State University, Bokkos. His research interests are in Media, Religion and Culture and Media and Conflict. He has published in several journals and books. He is also a member of Media, Religion and Culture (an international study group) as well as Signis World.

Festus Idoko is the former deputy editor of Nigeria Theatre Journal, and now Co-editor of two forthcoming festschrifts on Professor John Illah and Professor Emmy Idegu. Teaches Screen writing, Playcreating and Children Theatre at the Department of Theatre and Film Arts, University of Jos. Married with four children. Resident in Jos.

Edward S. Ossai, PhD, is a graduate of Ahmadu Bello University, Zaria and University of Jos, Nigeria. He had a stint as the Head of Studies and Assistant Director, Academic Planning at the National Film Institute, Jos, and tough film related courses at NTA, Television College, Jos, University of Jos Continuing Education Centre, Jos, Nasarawa State University, Keffi and National Institute for Cultural Orientation, Abuja. He is presently the Head, Theatre Arts Department, Plateau State University, Bokkos. Dr. Edward S. Ossai has over thirty short films, documentary films and feature length films to his credit. He has also published articles and chapters in journals and books. Dr. Edward S.Ossai likes reading, watching movies and travelling.

Index

N

O

P

R

S

T

U

W

www.ingramcontent.com/pod-product-compliance
Lightning Source LLC
LaVergne TN
LVHW061535290926
849162LV00004B/796